D0140940

Differential Equations

Differential Equations

A MODERN APPROACH

Harry Hochstadt

POLYTECHNIC INSTITUTE OF NEW YORK

DOVER PUBLICATIONS, INC.
NEW YORK

Published in Canada by General Publishing Company, Ltd., 30 Lesmill Road, Don Mills, Toronto, Ontario.
Published in the United Kingdom by Constable and Company, Ltd., 10 Orange Street, London WC 2.

This Dover edition, first published in 1975, is an unabridged republication, with minor corrections, of the work originally published by Holt, Rinehart and Winston, Inc., New York, in 1964.

International Standard Book Number: 0-486-61941-9
Library of Congress Catalog Card Number: 75-2569

Manufactured in the United States of America
Dover Publications, Inc.
180 Varick Street
New York, N.Y. 10014

Preface

In many ways the teaching of differential equations has not changed for several generations. The traditional courses still devote much time to those special cases where closed-form solutions can be found. This generally involves memorizing a variety of tricks and substitutions and where they apply.

In the past years the demands of applications have been becoming much stronger. Whereas, for example, until recently most engineers could get by with linear differential equations with constant coefficients, today they are much more concerned with time variable systems, which often are also nonlinear. The field of differential equations from the mathematician's viewpoint has in some ways kept up with these demands, but all too often this material is reserved for the more advanced texts and graduate courses.

The aim of this book is to present a modern viewpoint on differential equations in a framework that requires no prerequisites other than a good calculus course. A prior elementary course on differential equations would probably be of use but is not absolutely necessary. Certain concepts from linear algebra are used throughout, but the necessary background material is given in Chapter 1. Present trends seem to indicate that before long every science and engineering student will be given some basic linear algebra course as part of his undergraduate training. Basically this book can be used for a second course, which might come as early as the junior or senior year in the case of mathematics majors or in graduate school in the case of other science and engineering majors. Some of the basic material was given to freshmen honor students who had not had any previous courses on the subject. This experience showed that a prior course is not necessary although it probably is useful in the case of most students.

The first part of the book deals with linear differential equations. The basic existence theorems are proved, equations with constant coefficients are treated in detail, and series methods are investigated in detail for the cases where the coefficients are either analytic or have Fuchsian singularities, and also expansions in parameters. Chapter 4 is devoted to second-order boundary value problems. The self-adjoint cases are treated in detail and the concept of operators acting on vector spaces is used. The general treatment of the concomitant integral equations is also applicable to higher-order boundary value problems. Chapter 5 deals with linear differential equations with periodic coefficients.

The last two chapters of the book deal with nonlinear systems. The topics treated are existence and uniqueness theorems, series methods, stability theory, including a brief introduction to Lyapunov's direct method, and periodic systems. Chapter 7 treats some of the special situations that arise in second-order nonlinear systems.

The author has received help in the preparation of this book from a number of sources. The initial impetus came in connection with the teaching of a freshmen section of the Unified Honors Program at the Polytechnic Institute of Brooklyn. This program was initiated under a grant from the Ford Foundation and some of the time devoted to the preparation of the book was sponsored by the program. In much of the proofreading and in particular in the preparation of exercises the author was helped by two of his students, Bruce Stephan and Denis Taneri, who participated in an Undergraduate Science, Education and Research Program sponsored by the National Science Foundation. The most valuable help was given by Professor George Bachman, who read the whole manuscript very critically and made many helpful suggestions and corrections.

Brooklyn, New York H. H.
September, 1963

Contents

I Linear Algebra

1.1 INTRODUCTION

The purpose of this chapter is to familiarize the reader with those aspects of linear algebra that will be of importance in subsequent chapters. It is therefore not intended to be a complete exposition of the subject. A prior acquaintance with the subject is certainly preferable; if this is not the case, a subsequent more detailed study is highly recommended. However, the present section will provide an initial introduction and will provide reference material during study of the remaining chapters. A knowledge of determinants will be presupposed.

1.2 FINITE DIMENSIONAL VECTOR SPACES

We shall postpone a formal definition of vector spaces to a later section. A column n-vector will be defined as an array of n numbers, which may be real or complex, and denoted as follows

$$X = \begin{pmatrix} x_1 \\ x_2 \\ \vdots \\ x_n \end{pmatrix}.$$

The number x_i will be known as the ith component of X. In a similar fashion we shall define a row n-vector as

$$X^T = (x_1\, x_2 \cdots x_n).$$

The superscript T is to be read as "transpose." Evidently with every column n-vector X we can associate a row n-vector X^T.

We shall now define an operation of addition of two vectors as follows: Let

$$X = \begin{pmatrix} x_1 \\ x_2 \\ \vdots \\ x_n \end{pmatrix}, \quad Y = \begin{pmatrix} y_1 \\ y_2 \\ \vdots \\ y_n \end{pmatrix}, \quad Z = \begin{pmatrix} x_1 + y_1 \\ x_2 + y_2 \\ \vdots \\ x_n + y_n \end{pmatrix}.$$

Then, by definition,

$$X + Y = Z.$$

This additive operation has the following five important properties:

(a) The set of all column n-vectors is closed under addition. That is, the sum of two column n-vectors is a column n-vector.

(b) The operation is commutative:

$$X + Y = Y + X.$$

(c) The operation is associative:

$$(W + X) + Y = W + (X + Y).$$

(d) There is a 0 vector. That is,

$$0 = \begin{pmatrix} 0 \\ 0 \\ \vdots \\ 0 \end{pmatrix},$$

and

$$X + 0 = X.$$

(e) There is an inverse vector

$$Y = \begin{pmatrix} -x_1 \\ -x_2 \\ \vdots \\ -x_n \end{pmatrix},$$

so that

$$X + Y = 0.$$

A second operation is scalar multiplication. That is, given any number, a real or complex, we define

$$a\begin{pmatrix} x_1 \\ x_2 \\ \vdots \\ x_n \end{pmatrix} = \begin{pmatrix} ax_1 \\ ax_2 \\ \vdots \\ ax_n \end{pmatrix}.$$

This operation has the following properties:

(a) The set of all column n-vector is closed under scalar multiplication. That is, if X is a column n-vector, so is aX.

(b) The following distributive laws hold:

$$(a + b)X = aX + bX,$$
$$a(X + Y) = aX + aY.$$

(c) The operation is associative; that is,

$$a(bX) = (ab)X.$$

(d) There is an identity; that is,

$$1 \cdot X = X.$$

We now come to the concept of linear dependence.

DEFINITION: A set of nonvanishing column n-vectors $X_1, X_2, \cdots, X_k$ will be called linearly dependent (LD) if a set of numbers $c_1, c_2, \cdots c_k$ exists, at least one of which does not vanish, such that

$$\sum_{i=1}^{k} c_i X_i = 0.$$

If no such numbers exist, the set will be called linearly independent (LI).

EXAMPLE 1. The set of 3-vectors

$$X_1 = \begin{pmatrix} 1 \\ 2 \\ 3 \end{pmatrix}, \quad X_2 = \begin{pmatrix} 5 \\ 10 \\ 15 \end{pmatrix}$$

is LD, since

$$5X_1 - X_2 = 0.$$

The set of 3-vectors

$$X_1 = \begin{pmatrix} 1 \\ 0 \\ 0 \end{pmatrix}, \quad X_2 = \begin{pmatrix} 0 \\ 1 \\ 0 \end{pmatrix}, \quad X_3 = \begin{pmatrix} 0 \\ 0 \\ 1 \end{pmatrix}$$

is LI, since from

$$c_1X_1 + c_2X_2 + c_3X_3 = \begin{pmatrix} c_1 \\ c_2 \\ c_3 \end{pmatrix}$$

we conclude that all c_i must vanish if the above sum is to vanish. It is easy to see that it is impossible to find more than n LI n-vectors. For, if that were possible, we could find $X_1, X_2, \cdots X_{n+1}$ and $c_1, c_2, \cdots c_{n+1}$, not all zero such that

$$\sum_{i=1}^{n+1} c_i X_i = 0.$$

We now use the notation

$$X_i = \begin{pmatrix} x_{1i} \\ x_{2i} \\ \vdots \\ x_{ni} \end{pmatrix}, \qquad i = 1, 2, \cdots, n+1$$

so that x_{li} represents the lth component of the ith vector. By rewriting the preceding equation in scalar form, we have

$$\sum_{i=1}^{n+1} c_i x_{li} = 0; \qquad l = 1, 2, \cdots, n.$$

This is a system of n equations in $n + 1$ unknowns for the quantities c_i. Such a system must have nonvanishing solutions. It follows that the given vectors $X_1, X_2, \cdots X_{n+1}$ must be LD.

However, we can always find a set of n LI n-vectors. For example, one can verify that

$$X_1 = \begin{pmatrix} 1 \\ 0 \\ \vdots \\ 0 \end{pmatrix}, \quad X_2 = \begin{pmatrix} 0 \\ 1 \\ 0 \\ \vdots \\ 0 \end{pmatrix}, \quad \cdots, \quad X_n = \begin{pmatrix} 0 \\ 0 \\ \vdots \\ 1 \end{pmatrix}$$

is LI.

DEFINITION: A set of n LI n-vectors is called a *basis*.

A necessary and sufficient condition for a set of n n-vectors to be LI is that the determinant, formed by taking as the kth column the kth vector, does not vanish. This follows from the condition that if the vectors $X_1, X_2 \cdots, X_n$ are LI, then we should be able to conclude from

$$\sum_{i=1}^{n} c_i X_i = 0$$

that all $c_i = 0$. If we rewrite this system as the scalar system,

$$\sum_{i=1}^{n} c_i x_{li} = 0, \qquad l = 1, 2, \cdots, n,$$

we see that this system has one solution; that is, all $c_i = 0$, if and only if its determinant does not vanish.

The advantage of having a basis is that any vector can be expressed as a linear combination of the basis vectors; that is, if $X_1, X_2, \cdots X_n$ is a basis, then any other X can be expressed in the form

$$X = \sum_{i=1}^{n} c_i X_i.$$

EXAMPLE 2. Select the basis

$$X_1 = \begin{pmatrix} 1 \\ 0 \\ 0 \end{pmatrix}, \qquad X_2 = \begin{pmatrix} 0 \\ 1 \\ 0 \end{pmatrix}, \qquad X_3 = \begin{pmatrix} 0 \\ 0 \\ 1 \end{pmatrix}.$$

Then

$$X = \begin{pmatrix} x_1 \\ x_2 \\ x_3 \end{pmatrix} = x_1 X_1 + x_2 X_2 + x_3 X_3.$$

One can define the dimension of a finite dimensional vector space as the maximum number of LI vectors necessary to form a basis. Although a basis is not unique, one can show that the dimension of a space is independent of the basis selected.

1.3 INNER PRODUCTS

In ordinary vector analysis one has another operation called dot, or inner product. If V_1 and V_2 are two vectors in a three-dimensional Euclidean space, then the inner product

$$V_1 \cdot V_2 = |V_1| \, |V_2| \cos \theta,$$

where $|V|$ denotes the length of a vector and θ is the angle between V_1 and V_2. When two vectors are perpendicular, then $\theta = \pi/2$, and

$$V_1 \cdot V_2 = 0.$$

In this case we say that the vectors are orthogonal. One can see that the inner product has the following three properties:

(1) $V_1 \cdot V_2 = V_2 \cdot V_1$.

(2) $(aV_1 + bV_2) \cdot V_3 = a(V_1 \cdot V_3) + b(V_2 \cdot V_3)$.

(3) $V \cdot V = |V|^2 > 0$, if and only if $V \neq 0$.

We shall now try to generalize this inner product to n-dimensional vector spaces. To do so, we furnish the following three properties as a definition of the inner product, to be denoted by (X, Y). (X, Y) will be a number associated with the vectors X and Y such that

(1) $(X, Y) = \overline{(Y, X)}$ (the bar over the vectors denotes the complex conjugate.)

(2) $(aX + bZ, Y) = a(X, Y) + b(Z, Y)$.

(3) $(X, X) > 0$, if and only if $X \neq 0$.

Inner products are not unique, but the following definition of an inner product is the one that will be adopted by us.

We can verify that in an n-dimensional space, the following definition of an inner product

$$(X, Y) = \sum_{i=1}^{n} x_i \bar{y}_i,$$

where x_i and y_i are the ith components of X and Y, satisfies all three properties. One can deduce a host of other properties, such as

$$(X, aY + bZ) = \bar{a}(X, Y) + \bar{b}(X, Z).$$

We shall call two n-vectors X and Y orthogonal, if

$$(X, Y) = 0.$$

A basis $X_1, X_2, \cdots, X_n$ with the property that

$$(X_i, X_j) = 0 \qquad \text{for } i \neq j$$

will be called an orthogonal basis. If in addition to that we have

$$(X_i, X_i) = 1, \qquad i = 1, 2, \cdots n,$$

we shall call the basis orthonormal. Such orthonormal bases are of great advantage in computations. If $X_1, X_2 \cdots X_n$ denotes an arbitrary basis, then we know that (given any other n-vector X) we can find coefficients $c_1, c_2, \cdots, c_n$ such that

$$X = \sum_{i=1}^{n} c_i X_i.$$

But the computation of these coefficients is a messy algebraic problem. If we operate with an orthonormal basis, however, the computation becomes trivial. We then have

$$(X, X_j) = \left(\sum_{i=1}^{n} c_i X_i, X_j\right) = \sum_{i=1}^{n} c_i (X_i, X_j) = c_j.$$

The first step follows from property (2) of inner products and the second from the orthonormality of the basis. Hence we have found that

$$c_j = (X, X_j),$$

and we have

$$X = \sum_{i=1}^{n} (X, X_i)\, X_i.$$

One can use the inner product to introduce a distance concept into these vector spaces. For real three-dimensional vectors,

$$(X, X)^{1/2} = (x_1^2 + x_2^2 + x_3^2)^{1/2},$$

which is, of course, the ordinary Euclidian length of the 3-vector X. The distance between two vectors X and Y will be denoted by $(X - Y, X - Y)^{1/2}$, and from property 3, it can vanish if and only if $X = Y$.

There is an important inequality known as Schwarz's inequality, namely,

$$|(X, Y)|^2 \leqslant (X, X)(Y, Y).$$

To prove it, we observe that we can always write Y in the form

$$Y = \lambda X + Z,$$

where Z is such that

$$(X, Z) = 0,$$

and λ is so selected that

$$(X, Y - \lambda X) = 0.$$

The preceding equation can be rewritten in the form

$$\bar{\lambda}(X, X) = (X, Y).$$

Unless $X = 0$, it has a solution. If $X = 0$, the Schwarz inequality is trivially correct. Once λ has been selected, it is obvious that

$$(X, Z) = (X, Y - \lambda X) = 0.$$

Now

$$(Y, Y) = (\lambda X + Z, \lambda X + Z) = |\lambda|^2 (X, X) + (Z, Z),$$

and

$$\begin{aligned} |(X, Y)|^2 &= |\lambda|^2 (X, X)^2 = [|\lambda|^2 (X, X)]\,(X, X) \\ &= [(Y, Y) - (Z, Z)]\,(X, X) \leqslant (Y, Y)(X, X). \end{aligned}$$

Evidently the inequality reduces to an equality if $(Z, Z) = 0$, which implies that Y and X are LD, since

$$Y = \lambda X.$$

1.4 LINEAR TRANSFORMATIONS AND MATRICES

A transformation $\mathcal{A}$ that, when applied to an n-vector X, produces another n-vector Y is called a linear transformation if it has the properties

(1) $(aX)\mathcal{A} = a(X\mathcal{A})$.

(2) $(X + Z)\mathcal{A} = X\mathcal{A} + Z\mathcal{A}$.

One can obtain concrete realizations of such linear transformations in terms of matrices. Suppose $X_1, X_2, \cdots X_n$ is basis. We can construct n new vectors $Y_1, Y_2, \cdots Y_n$ defined by

$$X_i\mathcal{A} = Y_i, \qquad i = 1, 2, \cdots n.$$

Since the X_i form a basis, it must be possible to express the Y_i in terms of the X_i. Then

$$X_i\mathcal{A} = \sum_{j=1}^{n} a_{ji}X_j, \qquad i = 1, 2, \cdots n.$$

The n^2 coefficients a_{ij} fully determine the linear transformations, and the square array

$$(a_{ij}) = \begin{pmatrix} a_{11} & a_{12} & \cdots & a_{1n} \\ a_{21} & a_{22} & & a_{2n} \\ \vdots & \vdots & & \vdots \\ a_{n1} & a_{n2} & & a_{nn} \end{pmatrix}$$

will be called an $n \times n$ matrix. If $\mathcal{A}$ and $\mathcal{B}$ are two different linear transformations, we can evidently form a new transformation, called their sum, as follows:

$$X(\mathcal{A} + \mathcal{B}) = X\mathcal{A} + X\mathcal{B}.$$

since

$$X_i\mathcal{A} = \sum_{j=1}^{n} a_{ji}X_j$$

and

$$X_i\mathcal{B} = \sum_{j=1}^{n} b_{ji}X_j,$$

$$X_i(\mathcal{A} + \mathcal{B}) = \sum_{j=1}^{n} (a_{ji} + b_{ji})\, X_j.$$

From the operation of addition of two transformations one can define addition of two $n \times n$ matrices as follows: Using the shorthand notation

$$A = (a_{ij}),$$

we then have

$$A + B = (a_{ij}) + (b_{ij}) = (a_{ij} + b_{ij}).$$

One can easily check that this additive operation is commutative and associative; that is,

$$A + B = B + A$$

and

$$(A + B) + C = A + (B + C).$$

One can define the 0 matrix as one whose elements are all zeros. Then

$$A + 0 = A$$

and the matrix

$$B = (-a_{ij})$$

is an additive inverse of A, so that

$$A + B = 0.$$

One can also define a product of two transformations:

$$X_i(\mathscr{B}\mathscr{A}) = (X_i\mathscr{B})\mathscr{A} = \left(\sum_{k=1}^{n} b_{ki}X_k\right)\mathscr{A} = \sum_{k=1}^{n} b_{ki}(X_k\mathscr{A})$$

$$= \sum_{k=1}^{n} b_{ki} \sum_{j=1}^{n} a_{jk}X_j = \sum_{j=1}^{n} \left(\sum_{k=1}^{n} a_{jk}b_{ki}\right) X_j .$$

For the multiplicative operation we find that

$$AB = \left(\sum_{k=1}^{n} a_{ik}b_{kj}\right).$$

From the definition of an inner product one can see that if the row vector R_i^T, given by the ith row of A, is

$$R_i^T = (a_{i1}a_{i2} \cdots a_{in})$$

and the column vector C_j, given by the jth column of B, is

$$C_j = \begin{pmatrix} b_{1j} \\ b_{2j} \\ \vdots \\ b_{nj} \end{pmatrix},$$

then the i, j element of AB can be represented by

$$\sum_{k=1}^{n} a_{ik} b_{kj} = (R_i^T, \bar{C}_j).$$

In analogy to matrix multiplication one often uses the following notation for inner products of vectors:

$$(X, Y) = X^T \bar{Y}.$$

This notation will be used in subsequent chapters.

It can be verified that matrix multiplication is associative; that is,

$$(AB)C = A(BC),$$

but it is not commutative, since in general

$$AB \neq BA.$$

EXAMPLE. Let

$$A = \begin{pmatrix} 1 & 2 \\ 3 & 4 \end{pmatrix}, \qquad B = \begin{pmatrix} 7 & 8 \\ 0 & 1 \end{pmatrix},$$

$$AB = \begin{pmatrix} 7 & 10 \\ 21 & 28 \end{pmatrix},$$

$$BA = \begin{pmatrix} 31 & 46 \\ 3 & 4 \end{pmatrix},$$

so that $AB \neq BA$.

There is an identity transformation I such that

$$IX = X$$

for all X.

The matrix to be associated with I is given by

$$I = \begin{pmatrix} 1 & 0 & 0 & \cdots & 0 \\ 0 & 1 & 0 & \cdots & 0 \\ 0 & 0 & 1 & \cdots & 0 \\ \vdots & \vdots & \vdots & & \vdots \\ 0 & 0 & 0 & \cdots & 1 \end{pmatrix}$$

By using the definition of multiplication, one sees that

$$IA = AI = A.$$

The next question that arises in a natural way involves multiplicative inverses. That is, given a matrix A, can we find a matrix B such that

$$AB = I.$$

This question can be answered easily if one returns to the linear transformation viewpoint. Suppose we have a basis $X_1, X_2, \cdots X_n$ and a linear transformation $\mathcal{A}$ such that $X_1\mathcal{A}, X_2\mathcal{A}, \cdots, X_n\mathcal{A}$ is again a basis. Then

$$Y_i = X_i\mathcal{A} = \sum_{j=1}^{n} a_{ji}X_j.$$

But, in view of the fact that $Y_1, Y_2, \cdots Y_n$ forms a basis, it must be possible to express all X_i in terms of Y_i so that

$$X_i = \sum_{j=1}^{n} b_{ji}Y_j.$$

Interpreting the latter as another linear transformation, we find that if X is any vector expressed in the basis $X_1, X_2, \cdots X_n$, then

$$X\mathcal{A} = Y$$

and

$$X = Y\mathcal{B};$$

but

$$X\mathcal{A} = Y\mathcal{B}\mathcal{A} = Y,$$

so that $Y\mathcal{B}\mathcal{A} = Y$ implies that $\mathcal{B}\mathcal{A} = I$. Similarly $Y\mathcal{B} = X\mathcal{A}\mathcal{B} = X$, so that

$$X\mathcal{A}\mathcal{B} = X$$

and $\mathcal{A}\mathcal{B} = I$. We see, therefore, that a necessary and sufficient condition for $\mathcal{A}$ having an inverse is that a basis be transformed into a basis. Furthermore we observe that a multiplication of $\mathcal{A}$ by its inverse is commutative.

There is a very convenient criterion for a matrix having an inverse. As seen above, we can write out the transformation equations in terms of the scalar equations

$$y_{ki} = \sum_{j=1} a_{ji}x_{kj}, \qquad i = 1, 2, \cdots n.$$

Here y_{ki} denotes the kth components of Y_i. It must be possible to solve this system for $x_{k1}, x_{k2}, \cdots x_{kn}$. But this, we know, can be done only if the determinant of the coefficients does not vanish. Then

$$\begin{vmatrix} a_{11} & a_{12} & \cdots & a_{1n} \\ a_{21} & a_{22} & \cdots & a_{2n} \\ \vdots & \vdots & & \vdots \\ a_{n1} & a_{n2} & & a_{nn} \end{vmatrix} \neq 0.$$

This determinant is denoted either by the symbols $|A|$ or $\det(A)$. One can easily check that if X is transformed into Y by $\mathcal{A}$, then

$$X\mathcal{A} = Y,$$

where

$$X = \begin{pmatrix} x_1 \\ x_2 \\ \vdots \\ x_n \end{pmatrix},$$

$$Y = \begin{pmatrix} y_1 \\ y_2 \\ \vdots \\ y_k \end{pmatrix}$$

and

$$y_i = \sum_{j=1}^{n} a_{ij}x_j.$$

The preceding equation yields in essence a definition of the multiplication of a vector by a matrix.

It should be observed that from the viewpoint of notation, it is convenient to denote the operation of $\mathcal{A}$ acting on X by $X\mathcal{A}$, but the multiplication of X by the matrix A by AX. Thus

$$A\begin{pmatrix} x_1 \\ x_2 \\ \vdots \\ x_n \end{pmatrix} = \begin{pmatrix} \sum_{i=1}^{n} a_{1i}x_i \\ \sum_{i=1}^{n} a_{2i}x_i \\ \vdots \\ \sum_{i=1}^{n} a_{ni}x_i \end{pmatrix}$$

With every matrix A we can associate a transposed matrix A^T, which is obtained by replacing the ith column in A by the ith row, for all i. That is, if

$$A = \begin{pmatrix} a_{11} & a_{12} & \cdots & a_{1n} \\ a_{21} & a_{22} & \cdots & a_{2n} \\ \vdots & \vdots & & \vdots \\ a_{n1} & a_{n2} & \cdots & a_{nn} \end{pmatrix},$$

then

$$A^T = \begin{pmatrix} a_{11} & a_{21} & \cdots & a_{n1} \\ a_{12} & a_{22} & \cdots & a_{n2} \\ \vdots & \vdots & & \vdots \\ a_{1n} & a_{2n} & \cdots & a_{nn} \end{pmatrix}.$$

In many applications one also introduces the concept of the hermitian transpose A^+, defined by

$$A^+ = \bar{A}^T,$$

where the bar in $\bar{A}^T$ denotes the complex conjugate of A^T; that is, the matrix obtained by replacing every component of the A^T by its complex conjugate. When A contains only real elements, then there is no difference between the transpose and the hermitian transpose.

If $A^+ = A$, we say that the matrix is hermitian. A real hermitian matrix is called symmetric.

Such matrices play an important role in many applications. For example, in many problems one deals with inner products of the type (AX, Y) and it is often desirable to shift the operation from X to Y. A simple calculation then shows that

$$(AX, Y) = (X, A^+Y),$$

and in particular when A is hermitian,

$$(AX, Y) = (X, AY),$$

so that as far as the inner product is concerned, it makes no difference whether the operation is applied to X or Y.

1.5 COMPUTATIONAL ASPECTS OF MATRIX OPERATIONS

This section will be devoted to a discussion of some techniques that are useful in computations involving matrices; in particular, finding inverses of matrices.

We shall first define three so-called elementary operations:

(1) Multiplication of a given row of a matrix by a scalar c

(2) Interchange of two rows of a matrix

(3) Subtracting a given row from another row

Next we show that each of these operations can be performed by multiplying a given matrix by suitably selected other, so-called, elementary matrices.

Operation 1. Let

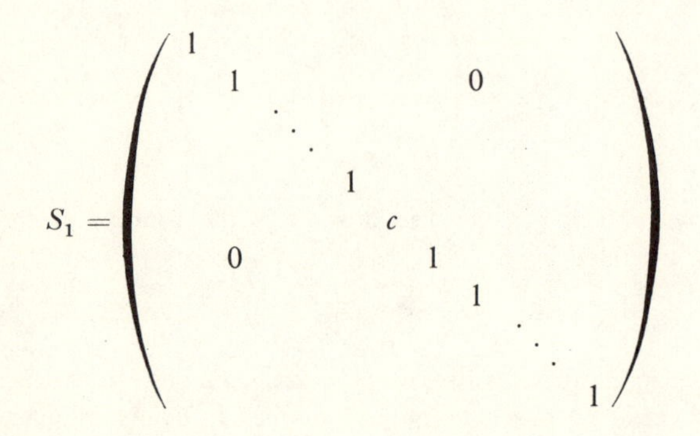

where all off-diagonal elements vanish and all elements on the main diagonal are equal to unity (with the exception of the ith element, which is equal to c). From the definition of matrix multiplication it follows that the matrix S_1A is a matrix that differs from A only in that all elements in the ith row are multiplied by c.

EXAMPLE 1.

$$\begin{pmatrix} 1 & 0 & 0 \\ 0 & c & 0 \\ 0 & 0 & 1 \end{pmatrix}\begin{pmatrix} a_{11} & a_{12} & a_{13} \\ a_{21} & a_{22} & a_{23} \\ a_{31} & a_{32} & a_{33} \end{pmatrix} = \begin{pmatrix} a_{11} & a_{12} & a_{13} \\ ca_{21} & ca_{22} & ca_{23} \\ a_{31} & a_{32} & a_{33} \end{pmatrix}.$$

From the fundamental rules of determinant operations we find that

$$|S_1| = c$$

and

$$|S_1A| = c|A|,$$

so that $|S_1A| = |S_1||A|$.

In other words, for this special case we have the rule that the determinant of the product of two matrices is the product of the determinant of the matrices.

Operation 2. To produce the second type of elementary operation, we create a matrix S_2 by starting with an identity matrix I and interchanging its ith and jth rows. Then S_2A differs from A only in that the ith and jth rows have been interchanged.

EXAMPLE 2.

$$\begin{pmatrix} 0 & 1 & 0 \\ 1 & 0 & 0 \\ 0 & 0 & 1 \end{pmatrix} \begin{pmatrix} a_{11} & a_{12} & a_{13} \\ a_{21} & a_{22} & a_{23} \\ a_{31} & a_{32} & a_{33} \end{pmatrix} = \begin{pmatrix} a_{21} & a_{22} & a_{23} \\ a_{11} & a_{12} & a_{13} \\ a_{31} & a_{32} & a_{33} \end{pmatrix}.$$

Again we see that, since

$$|S_2| = -1$$

and

$$|S_2A| = -|A|,$$
$$|S_2A| = |S_2||A|.$$

Operation 3. To produce the third type of operation, we create a matrix S_3. From the ith row we wish to subtract the jth row. The matrix S_3 appears as follows:

$$S_3 = \begin{pmatrix} 1 & 0 & 0 & \cdots & \cdots & \cdots & \cdots & 0 \\ 0 & 1 & 0 & \cdots & \cdots & \cdots & \cdots & 0 \\ \vdots & & & & & & & \\ 0 & 0 & 0 & \cdots & 01\,0 & 0\text{-}10 & \cdots & 0 \\ 0 & 0 & 0 & \cdots & \cdots & \cdots & \cdots & 1 \end{pmatrix} \begin{matrix} \\ \\ \\ \leftarrow i\text{th row} \\ \\ \end{matrix}$$

(with arrows marking the ith column and jth column)

Then S_3A is a matrix differing from A only in that the jth row was subtracted from the ith row.

EXAMPLE 3.

$$\begin{pmatrix} 1 & 0 & 0 \\ 0 & 1 & 0 \\ -1 & 0 & 1 \end{pmatrix} \begin{pmatrix} a_{11} & a_{12} & a_{13} \\ a_{21} & a_{22} & a_{23} \\ a_{31} & a_{32} & a_{33} \end{pmatrix} = \begin{pmatrix} a_{11} & a_{12} & a_{13} \\ a_{21} & a_{22} & a_{23} \\ a_{31}-a_{11} & a_{32}-a_{12} & a_{33}-a_{13} \end{pmatrix}.$$

Here we find that

$$|S_3| = 1$$
$$|S_3A| = |A|,$$

so that

$$|S_3A| = |S_3||A|.$$

A triangular matrix is a matrix in which all elements on one side of the main diagonal are identically zero. For example, the following is typical:

$$B = \begin{pmatrix} b_{11} & b_{12} & b_{13} & \cdots & b_{1n} \\ 0 & b_{22} & b_{23} & \cdots & b_{2n} \\ 0 & 0 & b_{33} & \cdots & b_{3n} \\ \vdots & \vdots & \vdots & & \vdots \\ 0 & 0 & 0 & \cdots & b_{nn} \end{pmatrix}.$$

One can state this analytically by saying

$$b_{ij} = 0 \qquad \text{for } i > j.$$

We shall now show that after a suitable number of elementary row operations on A, a triangular matrix can be produced.

First we consider the case where not all elements in the first column vanish. Then either a_{11} vanishes or does not vanish. If it does, but $a_{k1} \neq 0$, one can interchange the first and kth rows so that a_{k1} falls into the position formerly held by a_{11}. Now, by a relabeling process, $a_{11} \neq 0$. Next, by a combination of operations of types (1) and (3), a_{21} can be made to vanish. One merely subtracts a_{21}/a_{11} times the first row from the second. By a repetition of such operations all other elements in the first column can be made to vanish. Then the first row can be multiplied by $1/a_{11}$. The new matrix now has a first column that has one of two forms

$$\begin{matrix} 0 \\ 0 \\ \vdots \\ 0 \end{matrix} \qquad \text{or} \qquad \begin{matrix} 1 \\ 0 \\ \vdots \\ 0 \end{matrix}$$

The first is obtained if all elements in the first column vanished to begin with; the second, if at least one element did not.

Next, either all elements in the second column below a_{12} vanish or at least one does not. In the latter case we can assume that $a_{22} \neq 0$; if one or more do not, we proceed as before and by suitable row interchanges find a new $a_{22} \neq 0$. Then, after more row operations, we produce a second column of the form

$$\begin{matrix} a_{12} \\ 0 \\ 0 \\ \vdots \\ 0 \end{matrix} \qquad \text{or} \qquad \begin{matrix} a_{12} \\ 1 \\ 0 \\ \vdots \\ 0 \end{matrix}$$

Evidently, after a sufficient number of such operations, the ith column is such that all $a_{ik} = 0$ for $k < i$ and $a_{ii} = 0$ or $a_{ii} = 1$.

The nonvanishing elements can appear only on and above the main diagonal; therefore this matrix is a triangular matrix.

EXAMPLE 4. Let

$$A = \begin{pmatrix} 1 & 1 & 1 \\ 1 & 2 & 4 \\ 1 & 3 & 9 \end{pmatrix}.$$

After subtracting the first row from the second row and then from the third row, one finds

$$\begin{pmatrix} 1 & 0 & 0 \\ -1 & 1 & 0 \\ -1 & 0 & 1 \end{pmatrix} \begin{pmatrix} 1 & 1 & 1 \\ 1 & 2 & 4 \\ 1 & 3 & 9 \end{pmatrix} = \begin{pmatrix} 1 & 1 & 1 \\ 0 & 1 & 3 \\ 0 & 2 & 8 \end{pmatrix}.$$

Next the second row will be subtracted from the first and double the second row will be subtracted from the third row:

$$\begin{pmatrix} 1 & -1 & 0 \\ 0 & 1 & 0 \\ 0 & -2 & 1 \end{pmatrix} \begin{pmatrix} 1 & 1 & 1 \\ 0 & 1 & 3 \\ 0 & 2 & 8 \end{pmatrix} = \begin{pmatrix} 1 & 0 & -2 \\ 0 & 1 & 3 \\ 0 & 0 & 2 \end{pmatrix}.$$

Finally the third row will be multiplied by 1/2; then twice the third row will be added to the first, and three times the last row will be subtracted from the second:

$$\begin{pmatrix} 1 & 0 & 1 \\ 0 & 1 & -\frac{3}{2} \\ 0 & 0 & \frac{1}{2} \end{pmatrix} \begin{pmatrix} 1 & 0 & -2 \\ 0 & 1 & 3 \\ 0 & 0 & 2 \end{pmatrix} = \begin{pmatrix} 1 & 0 & 0 \\ 0 & 1 & 0 \\ 0 & 0 & 1 \end{pmatrix}.$$

By combining all these elementary row operations, we have

$$\begin{pmatrix} 1 & 0 & 1 \\ 0 & 1 & -\frac{3}{2} \\ 0 & 0 & \frac{1}{2} \end{pmatrix} \begin{pmatrix} 1 & -1 & 0 \\ 0 & 1 & 0 \\ 0 & -2 & 1 \end{pmatrix} \begin{pmatrix} 1 & 0 & 0 \\ -1 & 1 & 0 \\ -1 & 0 & 1 \end{pmatrix} = \begin{pmatrix} 3 & -3 & 1 \\ -\frac{5}{2} & 4 & -\frac{3}{2} \\ \frac{1}{2} & -1 & \frac{1}{2} \end{pmatrix}.$$

Finally we have

$$\begin{pmatrix} 3 & -3 & 1 \\ -\frac{5}{2} & 4 & -\frac{3}{2} \\ \frac{1}{2} & -1 & \frac{1}{2} \end{pmatrix} \begin{pmatrix} 1 & 1 & 1 \\ 1 & 2 & 4 \\ 1 & 3 & 9 \end{pmatrix} = \begin{pmatrix} 1 & 0 & 0 \\ 0 & 1 & 0 \\ 0 & 0 & 1 \end{pmatrix}.$$

We see, therefore, that we can find a matrix S, a product of elementary matrices of type S_1, S_2, and S_3 such that SA is triangular and elements in the diagonal are 0 or 1. Then we have either

$$|SA| = 0 \qquad \text{or} \qquad |SA| = 1.$$

In the first case we find that

$$|SA| = |S||A| = 0,$$

but $|S| \neq 0$, so that $|A| = 0$. In the second case all elements in the main diagonal are 1. By repeated row operations we can reduce all elements above the main diagonal to 0 so that $SA = I$. The preceding example illustrates this case. Here again we see that if $|A| \neq 0$, an inverse matrix must exist, but the foregoing procedure is a constructive device for finding the inverse. In the sequel the inverse of A will be designated by A^{-1}.

Another conclusion we can draw from our construction is that, in general,

$$|AB| = |A||B|.$$

First, if A is nonsingular (that is, $|A| \neq 0$), we have a matrix S, a product of elementary matrices such that

$$SA = I.$$

Then

$$|SA| = |S||A| = 1$$

and

$$|S| = \frac{1}{|A|}.$$

Then, if

$$AB = C,$$
$$SAB = IB = B = SC$$

and

$$|B| = |SC| = |S||C| = \frac{1}{|A|}|AB|,$$

and we have

$$|AB| = |A||B|.$$

Second, if $|A| = 0$, there must exist a matrix S such that SA has at least one vanishing diagonal element. By suitable row operations we can make that one the a_{nn} element. But in that case the entire last row of SA contains only zeros. Then, if

$$AB = C,$$
$$SAB = SC,$$

and $|SAB| = 0$, since if SA contains a zero row, so will SAB. Then

$$|SAB| = |S||AB| = 0,$$

and since $|S| \neq 0$, we find that

$$|AB| = 0,$$

and again we have

$$|AB| = |A||B|.$$

These operations will now be applied to the study of solutions of algebraic equations. We see that the system

$$\begin{aligned}
a_{11}x_1 + a_{12}x_2 + \cdots + a_{1n}x_n &= b_1, \\
a_{21}x_1 + a_{22}x_2 + \cdots + a_{2n}x_n &= b_2, \\
\vdots \qquad \vdots \qquad\qquad \vdots \quad &\quad \vdots \\
a_{n1}x_1 + a_{n2}x_2 + \cdots + a_{nn}x_n &= b_n,
\end{aligned}$$

can be rewritten in the form

$$AX = B,$$

where X and B are n-vectors and A is the matrix formed with the coefficients. If A has an inverse, we can find an S such that

$$SAX = X = SB,$$

so that we have a solution X; furthermore the solution must be unique.

If, however, A is singular, we find an S such that SA is triangular and has at least one zero, in the main diagonal:

$$SAX = SB.$$

In this case the system has either no solution or an infinity of solutions. The following examples will illustrate this point.

Example 5. Consider the system

$$\begin{aligned}
x_1 + x_2 + x_3 &= b_1, \\
x_1 + 2x_2 + 4x_3 &= b_2, \\
x_1 + \tfrac{3}{2}x_2 + \tfrac{5}{2}x_3 &= b_3.
\end{aligned}$$

By writing this system in matrix form, we have

$$\begin{pmatrix} 1 & 1 & 1 \\ 1 & 2 & 4 \\ 1 & \frac{3}{2} & \frac{5}{2} \end{pmatrix} \begin{pmatrix} x_1 \\ x_2 \\ x_3 \end{pmatrix} = \begin{pmatrix} b_1 \\ b_2 \\ b_3 \end{pmatrix}.$$

By multiplying both sides by the S matrix,

$$S = \begin{pmatrix} 2 & -1 & 0 \\ -1 & 1 & 0 \\ -\frac{1}{2} & -\frac{1}{2} & 1 \end{pmatrix},$$

we have

$$\begin{pmatrix} 1 & 0 & -2 \\ 0 & 1 & 3 \\ 0 & 0 & 0 \end{pmatrix} \begin{pmatrix} x_1 \\ x_2 \\ x_3 \end{pmatrix} = \begin{pmatrix} 2b_1 - b_2 \\ -b_1 + b_2 \\ -\frac{1}{2}b_1 - \frac{1}{2}b_2 + b_3 \end{pmatrix}.$$

Unless the compatibility condition

$$-\tfrac{1}{2}b_1 - \tfrac{1}{2}b_2 + b_3 = 0$$

is satisfied, no solution exists. If the system satisfies the compatibility condition, one can rewrite the system as one in two unknowns. Then

$$\begin{aligned} x_1 &= 2b_1 - b_2 + 2x_3, \\ x_2 &= -b_1 + b_2 - 3x_3. \end{aligned}$$

Now x_3 is chosen arbitrarily and then x_1 and x_2 are determined; but one has as many solutions as there are choices for x_3, namely, an infinity.

One can conclude in a similar fashion that the system

$$AX = B,$$

after being put into the triangular form

$$SAX = SB,$$

will have a unique solution if SA is nonsingular. If SA is singular, then at least one row consists only of zeros, and the system must satisfy a suitable compatibility condition in order for a solution to exist; but such solutions are no longer unique.

1.6 EIGENVALUES AND EIGENVECTORS

In the computation of terms like AX, we saw that once a basis has been selected and once the effect of multiplication of all the X_i by A has been defined, it is possible to determine AX. For example, if

$$AX_i = \sum_{j=1}^{n} a_{ji} X_j, \qquad i = 1, 2, \cdots, n$$

and

$$X = \sum_{i=1}^{n} a_i X_i,$$

then

$$AX = \sum_{i=1}^{n} a_i AX_i = \sum_{i=1}^{n} a_i \sum_{j=1}^{n} a_{ji} X_j = \sum_{j=1}^{n} \left(\sum_{j=1}^{n} a_i a_{ji} \right) X_j.$$

We are therefore led to ask the question whether or not some bases are more convenient than others. It seems plausible that the simplest type of basis is such that

$$AX_i = \lambda_i X_i,$$

where λ is a scalar. In others words, the only effect A has when multiplied into X_i is to change its length by a factor λ_i. Then, of course, if

$$X = \sum_{i=1}^{n} a_i X_i,$$

$$AX = \sum_{i=1}^{n} a_i \lambda_i X_i.$$

We shall now show that, under suitable conditions, this effect can always be achieved. Before investigating whether or not such a basis can be found, we shall show that at least one vector X can be found such that

$$AX = \lambda X,$$

where λ is a suitable scalar. We then rewrite the equation in the form

$$(A - \lambda I) X = 0.$$

This is a homogeneous system of n equations in the n unknown components of X. It will have either only the solution $X = 0$ or an infinity of solutions, depending on whether the determinant of the system does not or does vanish. Since we are interested in nonvanishing solutions, we require that

$$|A - \lambda I| = \begin{vmatrix} a_{11} - \lambda & a_{12} & \cdots & a_{1n} \\ a_{21} & a_{22} - \lambda & \cdots & a_{2n} \\ \vdots & \vdots & & \vdots \\ a_{n1} & a_{n2} & \cdots & a_{nn} - \lambda \end{vmatrix} = 0.$$

This leads evidently to an algebraic equation of degree n for λ. This equation must, according to the fundamental theorem of algebra, have n roots. Once a λ has been selected in this way, we can find a nonvanishing X such that

$$AX = \lambda X.$$

We now suppose that the above determinant equation has n distinct roots: $\lambda_1, \lambda_2, \ldots, \lambda_n$. Corresponding to them we find n vectors $X_1, X_2, \cdots X_n$ such that

$$AX_i = \lambda_i X_i, \qquad i = 1, 2, \cdots, n.$$

The λ_i are known as eigenvalues, characteristic values, or proper values. The X_i are known as eigenvectors, characteristic vectors, or proper vectors. We shall now show that these eigenvectors are linearly independent so that they form a basis. The proof will be by contradiction. We suppose that the X_i are linearly dependent so that a set of scalars c_i exist such that

$$\sum_{i=1}^{n} c_i X_i = 0,$$

and at least one of the c_i does not vanish. Without loss of generality we can suppose that $c_1 \neq 0$. Then we consider the matrix

$$P = (A - \lambda_2 I)(A - \lambda_3 I) \cdots (A - \lambda_n I)$$

and observe that

$$\begin{aligned} PX_n &= (A - \lambda_2 I) \cdots (A - \lambda_{n-1} I)(A - \lambda_n I) X_n \\ &= (A - \lambda_2 I) \cdots (A - \lambda_{n-1} I)(AX_n - \lambda_n X_n) \\ &= 0, \end{aligned}$$

since the last factor must vanish by the definition of X_n. Similarly,

$$PX_k = 0, \qquad k = 2, 3, \cdots n.$$

But

$$\begin{aligned} PX_1 &= (A - \lambda_2 I) \cdots (A - \lambda_{n-1} I)(AX_1 - \lambda_n X_1) \\ &= (A - \lambda_2 I) \cdots (A - \lambda_{n-1} I)(\lambda_1 - \lambda_n) X_1 \\ &= (A - \lambda_2 I) \cdots (AX_1 - \lambda_{n-1} X_1)(\lambda_1 - \lambda_n) \\ &= (A - \lambda_2 I) \cdots (\lambda_1 - \lambda_{n-1})(\lambda_1 - \lambda_n) X_1 \\ &= \cdots \\ &= (\lambda_1 - \lambda_2)(\lambda_1 - \lambda_3) \cdots (\lambda_1 - \lambda_n) X_1. \end{aligned}$$

Next we see that

$$\begin{aligned} P \sum_{i=1}^{n} c_i X_i = \sum_{i=1}^{n} c_i P X_i &= c_1 P X_1 \\ &= c_1(\lambda_1 - \lambda_2) \cdots (\lambda_1 - \lambda_n) X_1 = 0. \end{aligned}$$

Since all λ_i are distinct, it must follow that $c_1 = 0$. It follows that all c_i must vanish, and this contradicts the assumption that the set of eigenvectors is linearly dependent.

The set of eigenvectors has, in addition to furnishing a particularly convenient basis, another application that will be important to us. We now construct a new matrix whose jth column is composed of the elements of X_j; that is,

$$T = \begin{pmatrix} x_{11} & x_{12} & \cdots & x_{1n} \\ x_{21} & x_{22} & \cdots & x_{2n} \\ \vdots & \vdots & & \vdots \\ x_{n1} & x_{n2} & & x_{nn} \end{pmatrix}.$$

Since the columns are linearly independent, T must have an inverse T^{-1}. We also have the simple identities

$$T\begin{pmatrix} 1 \\ 0 \\ 0 \\ \vdots \\ 0 \end{pmatrix} = X_1, \qquad T\begin{pmatrix} 0 \\ 1 \\ 0 \\ \vdots \\ 0 \end{pmatrix} = X_2 \cdots .$$

If one introduces the Kronecker δ_{ij} defined by

$$\begin{aligned} \delta_{ij} &= 0, \qquad i \neq j \\ &= 1, \qquad i = j, \end{aligned}$$

we can write in general that

$$T\begin{pmatrix} \delta_{1i} \\ \delta_{2i} \\ \vdots \\ \delta_{ni} \end{pmatrix} = X_i \qquad \text{and} \qquad T^{-1}X_i = \begin{pmatrix} \delta_{1i} \\ \delta_{2i} \\ \vdots \\ \delta_{ni} \end{pmatrix}.$$

Now we compute

$$T^{-1}AT\begin{pmatrix} \delta_{1i} \\ \delta_{2i} \\ \vdots \\ \delta_{ni} \end{pmatrix} = T^{-1}AX_i = T^{-1}\lambda_i X_i = \lambda_i T^{-1}X_i = \lambda_i\begin{pmatrix} \delta_{1i} \\ \delta_{2i} \\ \vdots \\ \delta_{ni} \end{pmatrix}.$$

We see, therefore, that the matrix $T^{-1}AT$ has the same eigenvalues as A (namely, $\lambda_1, \lambda_2, \cdots, \lambda_n$) and has as eigenvectors:

$$\begin{pmatrix} 1 \\ 0 \\ 0 \\ \vdots \\ 0 \end{pmatrix}, \begin{pmatrix} 0 \\ 1 \\ 0 \\ \vdots \\ 0 \end{pmatrix} \cdots \begin{pmatrix} \delta_{1i} \\ \delta_{2i} \\ \cdot \\ \vdots \\ \delta_{ni} \end{pmatrix} \cdots \begin{pmatrix} 0 \\ 0 \\ 0 \\ \vdots \\ 1 \end{pmatrix}.$$

Evidently the diagonal matrix

$$D = \begin{pmatrix} \lambda_1 & & & 0 \\ & \lambda_2 & & \\ & & \ddots & \\ 0 & & & \lambda_n \end{pmatrix}$$

has the same eigenvalues and eigenvectors. One can show that two matrices with the same eigenvalues and eigenvectors, since they define the same linear transformation, must be identical, so that

$$T^{-1}AT = D.$$

EXAMPLE.

$$A = \begin{pmatrix} 0 & 1 & 0 \\ 0 & 0 & 1 \\ 6 & -11 & 6 \end{pmatrix}$$

and

$$|A - \lambda I| = \begin{vmatrix} -\lambda & 1 & 0 \\ 0 & -\lambda & 1 \\ 6 & -11 & 6-\lambda \end{vmatrix} = -\lambda^3 + 6\lambda^2 - 11\lambda + 6$$

$$= -(\lambda - 1)(\lambda - 2)(\lambda - 3).$$

The eigenvalues are distinct, so that one can find three linearly independent eigenvectors X_1, X_2, X_3 as follows: For $\lambda = 1$, we have

$$(A - I)X_1 = \begin{pmatrix} -1 & 1 & 0 \\ 0 & -1 & 1 \\ 6 & -11 & 5 \end{pmatrix} \begin{pmatrix} x_{11} \\ x_{21} \\ x_{31} \end{pmatrix} = \begin{pmatrix} -x_{11} + x_{21} \\ -x_{21} + x_{31} \\ 6x_{11} - 11x_{21} + 5x_{31} \end{pmatrix} = 0.$$

From the first two equations one has

$$x_{11} = x_{21} = x_{31},$$

and the third is identically satisfied by this set of solutions. To find a suitable X_1, one selects from these solutions any one that does not vanish; say,

$$x_{11} = x_{21} = x_{31} = 1.$$

In this manner one can find three eigenvectors:

$$X_1 = \begin{pmatrix} 1 \\ 1 \\ 1 \end{pmatrix}, \quad X_2 = \begin{pmatrix} 1 \\ 2 \\ 4 \end{pmatrix}, \quad X_3 = \begin{pmatrix} 1 \\ 3 \\ 9 \end{pmatrix}$$

so that

$$AX_1 = X_1, \quad AX_2 = 2X_2, \quad AX_3 = 3X_3.$$

The matrix

$$T = \begin{pmatrix} 1 & 1 & 1 \\ 1 & 2 & 3 \\ 1 & 4 & 9 \end{pmatrix}$$

has the inverse

$$T^{-1} = \begin{pmatrix} 3 & -\frac{5}{2} & \frac{1}{2} \\ -3 & 4 & -1 \\ 1 & -\frac{3}{2} & \frac{1}{2} \end{pmatrix}$$

and a direct calculation shows that

$$T^{-1}AT = \begin{pmatrix} 1 & 0 & 0 \\ 0 & 2 & 0 \\ 0 & 0 & 3 \end{pmatrix}.$$

The case in which multiple eigenvalues occur will be discussed in the next chapter in connection with Laplace transform methods.

We shall now restrict ourselves to hermitian matrices. In this case, as we shall show, all eigenvalues are real. The proof is by contradiction. Suppose

$$AX = \lambda X,$$

where λ is complex. By taking complex conjugates of the above, we have

$$\bar{A}\bar{X} = \bar{\lambda}\bar{X}.$$

This shows that $\bar{\lambda}$ is an eigenvalue of $\bar{A}$ and has eigenvector $\bar{X}$. We now take the inner product of the first equation with X and the product of the second with $\bar{X}$, so that

$$(AX, X) = (\lambda X, X) = \lambda(X, X),$$
$$(\bar{A}\bar{X}, \bar{X}) = (\bar{\lambda}\bar{X}, \bar{X}) = \bar{\lambda}(\bar{X}, \bar{X}).$$

We note that (X, X) is real, and it follows that

$$(X, X) = \overline{(X, X)} = (\bar{X}, \bar{X}).$$

From the definition of inner product we have

$$(\bar{A}\bar{X}, \bar{X}) = \overline{(\bar{X}, \bar{A}\bar{X})} = (X, AX) = (A^{+}X, X) = (AX, X),$$

the last expression following from the fact that A is hermitian. Therefore

$$\lambda(X, X) = \bar{\lambda}(\bar{X}, \bar{X}),$$

but since $(X, X) > 0$, we see that

$$\lambda = \bar{\lambda},$$

so that λ is certainly real. We can also show that all eigenvectors are orthogonal. Suppose

$$AX_1 = \lambda_1 X_1,$$
$$AX_2 = \lambda_2 X_2.$$

Then

$$(AX_1, X_2) = (\lambda_1 X_1, X_2) = \lambda_1(X_1, X_2),$$

but since A is hermitian,

$$(AX_1, X_2) = (X_1, AX_2) = (X_1, \lambda_2 X_2) = \lambda_2(X_1, X_2).$$

Therefore

$$\lambda_1(X_1, X_2) = \lambda_2(X_1, X_2),$$

and since all λ_i were supposed to be distinct, it follows that

$$(X_1, X_2) = 0.$$

In many of the subsequent applications we shall be dealing with vectors and matrices whose components are functions of some variable, say, z. We shall then write $A(z)$ and $X(z)$, or just A and X, when it is clear from the context that they are functions. As functions they have properties very similar to scalar functions. When dealing with differentiable functions, one can easily deduce a number of formulas corresponding to well-known formulas in the case of scalar functions. We now define these operations as follows:

$$\frac{d}{dz} X(z) = \frac{d}{dz} \begin{pmatrix} x_1 & (z) \\ x_2 & (z) \\ & \vdots \\ x_n & (z) \end{pmatrix} = \begin{pmatrix} \frac{d}{dz} & x_1 & (z) \\ \frac{d}{dz} & x_2 & (z) \\ & \vdots & \\ \frac{d}{dz} & x_n & (z) \end{pmatrix},$$

$$\frac{d}{dz} A(z) = \frac{d}{dz} (a_{ij}(z)) = \left(\frac{d}{dz} a_{ij}(z)\right).$$

From these one can derive the formulas

$$\frac{d}{dz} [A(z) + B(z)] = \frac{d}{dz} A(z) + \frac{d}{dz} B(z),$$

$$\frac{d}{dz} [A(z)B(z)] = \left[\frac{d}{dz} A(z)\right] B(z) + A(z) \left[\frac{d}{dz} B(z)\right],$$

$$\frac{d}{dz} [A(z)X(z)] = \left[\frac{d}{dz} A(z)\right] X(z) + A(z) \frac{d}{dz} X(z),$$

$$\frac{d}{dz} [(X(z), Y(z))] = \left(\frac{d}{dz} X(z), Y(z)\right) + \left(X(z), \frac{d}{dz} Y(z)\right).$$

1.7 NORMS OF MATRICES

In dealing with ordinary numbers, we often refer to their absolute value or magnitude. This absolute value has the following three properties:

(1) $|x - y| \geqslant 0$, and vanishes if and only if $x = y$.

(2) $|x - y| = |y - x|$.

(3) $|x - z| \leqslant |x - y| + |y - z|$.

Such a function is often known as a distance function, since $|x - y|$ denotes the distance between the points x and y on a Euclidian line. One can define such distance functions in vector spaces because the function

$$|X - Y| = (X - Y, X - Y)^{1/2} = \left(\sum_{i=1}^{n}(x_i - y_i)^2\right)^{1/2}$$

for real vectors X and Y can easily be shown to have properties (1), (2), and (3). The first two are evident by inspection. To prove (3), we proceed as follows: Evidently we have

$$0 \leqslant \sum_{\substack{i,j=1\\ i\neq j}}^{n} (a_i b_j - a_j b_i)^2,$$

from which one obtains by a rearrangement of terms,

$$2 \sum_{\substack{i,j=1\\ i\neq j}}^{n} a_i b_i a_j b_j \leqslant 2 \sum_{\substack{i,j=1\\ i\neq j}}^{n} a_i^2 b_j^2.$$

By adding $\sum_1^n a_i^2 b_i^2$ to both sides, one has

$$\left(\sum_{i=1}^{n} a_i b_i\right)^2 \leqslant \left(\sum_{i=1}^{n} a_i^2\right)\left(\sum_{i=1}^{n} b_i^2\right).$$

After taking the square root of both sides, doubling, and adding

$$\sum_{i=1}^{n} a_i^2 + \sum_{i=1}^{n} b_i^2$$

one obtains

$$\left[\sum_{i=1}^{n} (a_i + b_i)^2\right]^{1/2} \leqslant \left[\sum_{i=1}^{n} a_i^2\right]^{1/2} + \left[\sum_{i=1}^{n} b_i^2\right]^{1/2}.$$

If one now sets

$$a_i = x_i - y_i,$$
$$b_i = y_i - z_i,$$
$$a_i + b_i = x_i - z_i,$$

one obtains the desired inequality, property (3). Such a distance function is not unique, since the following function, for example,

$$\| X - Y \| = \sum_{i=1}^{n} | x_i - y_i |$$

also satisfies properties (1), (2), and (3). However, the previously defined function $| X - Y |$ has the obvious interpretation of Euclidian distance, whereas the above function $\| X - Y \|$ has no obvious geometrical meaning.

It is possible to define such a function for matrices as well:

$$\| A - B \| = \sum_{i,j=1}^{n} | a_{ij} - b_{ij} |.$$

Since the right side is a sum of ordinary absolute values, we see that this function has properties (1), (2), and (3), and we also see that

$$\begin{aligned} \| AB \| &= \sum_{i,j=1}^{n} \left| \sum_{k=1}^{n} a_{ik} b_{kj} \right| \\ &\leqslant \sum_{i,j,k=1,}^{n} | a_{ik} | \, | b_{kj} | \leqslant \sum_{i,j,k,l=1}^{n} | a_{ik} | \, | b_{lj} | \\ &= \left(\sum_{i,k=1}^{n} | a_{ik} | \right) \left(\sum_{l,j=1}^{n} | b_{lj} | \right) = \| A \| \, \| B \|. \end{aligned}$$

The first inequality follows from property (3) for ordinary absolute values; namely,

$$| x + y | \leqslant | x | + | y |,$$

and by a simple extension,

$$| x_1 + x_2 + \cdots + x_n | \leqslant | x_1 | + | x_2 | + \cdots + | x_n |.$$

The second inequality follows from the fact that more nonnegative terms have been introduced on the right.

The function $\| A \|$ is known as a norm. Such a norm is of great use in the study of the convergence of infinite processes. Thus, for example, in considering an infinite series of matrices, say,

$$\sum_{n=0}^{\infty} A_n,$$

one would like to be able to decide whether or not, and in what sense, such a series can be said to converge. One can easily generalize the notion of convergence and say that the above series converges if the partial sums

$$S_n = \sum_0^n A_k$$

have a limit as n approaches infinity. If we let

$$\| A_k \| = u_k,$$

then

$$\| S_n \| \leqslant \sum_0^n u_k;$$

and if the infinite series

$$\sum_0^\infty u_k$$

converges, S_n will have a limit. A simple example is the following: The series

$$\sum_0^\infty A^n$$

converges, provided

$$\| A \| < 1.$$

The above test is a test for absolute convergence, which of course in turn implies convergence.

1.8 LINEAR VECTOR SPACES

DEFINITION: A set S of elements $u, v, w, \cdots$ is said to be a linear vector space (also called linear space and vector space) if the following conditions are satisfied:

(1) For any two elements u, v in S there exists a uniquely defined third element $w = u + v$ in S, called their sum, such that

(a) $u + v = v + u$.

(b) $u + (v + w) = (u + v) + w$.

(c) there exists an element 0 having the property that $u + 0 = u$ for all u in S.

(d) For every u in S there exists an element $-u$ such that $u + (-u) = 0$.

(2) For an ordinary scalar α and every element u in S there is defined the product αu such that

(a) $\alpha(\beta u) = (\alpha\beta)\, u$.

(b) $1 \cdot u = u$.

(3) These operations of addition and scalar multiplication are related in the following way:

(a) $(\alpha + \beta)\, u = \alpha u + \beta u$.

(b) $\alpha(u + v) = \alpha u + \alpha v$.

(4) A linear space is called a normed linear space if one can introduce for each u in S a nonnegative number $\| u \|$, called the norm of u, with the property that

(a) $\| u \| = 0$ if and only if $u = 0$.

(b) $\| \alpha u \| = | \alpha | \, \| u \|$.

(c) $\| u + v \| \leqslant \| u \| + \| v \|$.

(5) We shall also suppose that one can define an inner product in the space, that is, a scalar (u, v) depending on the two elements u an v such that

(a) $(u, v) = \overline{(v, u)}$.

(b) $(\alpha u + \beta v, w) = \alpha(u, w) + \beta(u, w)$.

(c) $(u, u) > 0$ if and only if $u \neq 0$.

A set of nonvanishing elements $u_1, u_2, \cdots u_n$ will be called linearly independent if the equality

$$\sum_{i=1}^{n} c_i u_i = 0,$$

where the c_i are scalars, can hold only if all c_i vanish.

A set of n linearly independent elements $u_1, u_2, \cdots u_n$, with the property that any set of $(n + 1)$ elements including these n elements is linearly dependent, is called a basis of the space. It follows that if $u, u_1, \cdots u_n$ are $(n + 1)$ such linearly dependent elements, then one can find scalars $c_0, c_1, \cdots c_u$ such that

$$c_0 u + c_1 u_1 + \cdots + c_n u_n = 0$$

If u does not vanish, c_0 also cannot vanish. Otherwise all c_i would have to vanish, since $u_1, u_2, \cdots u_n$ are linearly independent. Then

$$u = -\frac{c_1}{c_0} u_1 - \frac{c_2}{c_0} u_2 - \cdots - \frac{c_n}{c_0} u_n$$

so that u can be expressed in terms of the basis elements. Evidently a basis is a coordinate system. Such a basis is not unique, but one can show that the number n depends only on the space. Then any set of n linearly independent elements forms a basis, and one calls n the dimension of the space.

In some of our applications we shall deal with infinite dimensional spaces. But in these applications it will develop that it will be possible to find a basis u_1, $u_2, \cdots u_n \cdots$ such that any element u can be expressed in the form

$$u = \sum_{i=1}^{\infty} c_i u_i,$$

but it will be necessary to show that such series converge.

A transformation, or operator A, that transforms any element u in S into some element v in S will be called a linear transformation if it has the property:

$$A(\alpha u + \beta v) = \alpha A u + \beta A v.$$

A transformation A^+ with the property

$$(u, Av) = (A^+u, v)$$

is called the *adjoint* of A, and A is called a *self-adjoint transformation* if

$$(u, Av) = (Au, v).$$

In dealing with vector spaces, we see that all transformations can be represented by matrix multiplication and hermitian transposes, and hermitian matrices play the roles of adjoints and self-adjoint transformations.

Scalars λ and elements u such that

$$Au = \lambda u$$

are known as eigenvalues and eigenelements.

In our future applications we shall deal with self-adjoint transformations. As in the case of hermitian matrices one can show that all eigenvalues of self-adjoint transformations must be real. In our applications it will be possible to find eigenelements u_i and eigenvalues λ_i, and for the u_i we have

$$(u_i, u_j) = 0, \qquad i \neq j,$$

and, as will be seen later, one can always normalize u_i so that

$$(u_i, u_i) = 1.$$

If the u_i form a basis for the space with the above properties, then it will be called an *orthonormal basis*. In this case, if

$$u = \sum c_i u_i,$$

then

$$(u, u_j) = \sum c_i (u_i, u_j) = c_j,$$

so that

$$u = \sum (u, u_i)\, u_i.$$

One other aspect will be touched on here in generality, namely, the question of the solvability of the equation

$$Au - \lambda u = v. \tag{1}$$

By assuming that A is self-adjoint and has the distinct eigenvalues λ_i and eigenelements u_i, which form an orthonormal basis in S, one can write v in the form

$$v = \sum c_i u_i, \tag{2}$$

and then one finds that

$$u = \sum \frac{c_i}{\lambda_i - \lambda} u_i. \tag{3}$$

To verify that this is the solution, we compute

$$\begin{aligned}(A - \lambda) u &= \sum \frac{c_i}{\lambda_i - \lambda} (A - \lambda) u_i \\ &= \sum \frac{c_i}{\lambda_i - \lambda} (\lambda_i - \lambda) u_i = \sum c_i u_i = v.\end{aligned}$$

Evidently, if λ is not an eigenvalue, the solution of (1) is given by (3). But if λ is an eigenvalue, say, λ_k, a solution may not exist. One sees that for a solution to exist, we must introduce the compatibility condition on v that $c_k = 0$ so that (3) exists. Then

$$u = \sum_{i \neq k} \frac{c_i}{\lambda_i - \lambda_k} u_i + \alpha u_k, \tag{4}$$

where α is arbitrary. Hence we see that if λ is an eigenvalue, we have either no solution or an infinity. This should be compared to the corresponding problem of solving linear algebraic equations with and without vanishing determinant.

These methods will be applied in Chapter 4 to the solution of boundary-value problems.

Problems

1. Show that the following sets of vectors are linearly dependent:

(a) $\begin{pmatrix} 4 \\ 2 \end{pmatrix}, \quad \begin{pmatrix} 2 \\ 1 \end{pmatrix}, \quad \begin{pmatrix} -1 \\ 3 \end{pmatrix}.$

(b) $\begin{pmatrix} -5 \\ 2 \\ 8 \\ -16 \end{pmatrix}, \quad \begin{pmatrix} 1 \\ 0 \\ 2 \\ 4 \end{pmatrix}, \quad \begin{pmatrix} 0 \\ 1 \\ 9 \\ 2 \end{pmatrix}.$

(c) $\begin{pmatrix} 4 \\ -9 \\ 11 \end{pmatrix}, \quad \begin{pmatrix} 2 \\ 1 \\ 1 \end{pmatrix}, \quad \begin{pmatrix} 3 \\ -4 \\ 6 \end{pmatrix}.$

2. Determine k such that the following vectors are linearly independent:

$$\begin{pmatrix}1\\2\\k\end{pmatrix},\quad \begin{pmatrix}k-1\\2\\1\end{pmatrix},\quad \begin{pmatrix}3\\4\\2\end{pmatrix}.$$

3. Prove that if three 3-vectors are linearly dependent, then they lie in a plane.

4. Express the vector $X = \begin{pmatrix}2\\1\\1\end{pmatrix}$ in terms of the following basis vectors:

$$X_1 = \begin{pmatrix}1\\1\\1\end{pmatrix},\quad X_2 = \begin{pmatrix}1\\0\\-1\end{pmatrix},\quad X_3 = \begin{pmatrix}0\\-1\\2\end{pmatrix}.$$

5. Which of the following sets of vectors forms a basis in a three-dimensional space?

(a) $\begin{pmatrix}3\\0\\2\end{pmatrix},\quad \begin{pmatrix}1\\1\\0\end{pmatrix},\quad \begin{pmatrix}1\\9\\7\end{pmatrix}.$

(b) $\begin{pmatrix}7\\0\\9\end{pmatrix},\quad \begin{pmatrix}3\\0\\1\end{pmatrix},\quad \begin{pmatrix}4\\0\\6\end{pmatrix}.$

(c) $\begin{pmatrix}4\\1\\2\end{pmatrix},\quad \begin{pmatrix}5\\2\\1\end{pmatrix},\quad \begin{pmatrix}1\\0\\1\end{pmatrix}.$

6. For which values of k will the following vectors form a basis in a three-dimensional space?

$$\begin{pmatrix}1\\0\\-2k\end{pmatrix},\quad \begin{pmatrix}1\\3\\-11\end{pmatrix},\quad \begin{pmatrix}2\\-k\\-7\end{pmatrix}.$$

7. Complete the following sets of vectors to obtain bases in a four-dimensional space:

(a) $\begin{pmatrix}1\\1\\-1\\1\end{pmatrix},\quad \begin{pmatrix}1\\0\\1\\1\end{pmatrix},\quad \begin{pmatrix}1\\2\\1\\1\end{pmatrix}.$

(b) $\begin{pmatrix}1\\3\\5\\2\end{pmatrix},\quad \begin{pmatrix}-1\\-3\\2\\5\end{pmatrix},\quad \begin{pmatrix}0\\0\\1\\1\end{pmatrix}.$

(c) $\begin{pmatrix}-3\\4\\6\\2\end{pmatrix},\quad \begin{pmatrix}3\\5\\1\\0\end{pmatrix},\quad \begin{pmatrix}0\\1\\2\\9\end{pmatrix}.$

8. Compute (X, Y) for the following:

(a) $X = \begin{pmatrix} 1 \\ i+1 \\ 2 \\ -i-2 \end{pmatrix}, \quad Y = \begin{pmatrix} -3 \\ 1 \\ i+2 \\ 0 \end{pmatrix}.$

(b) $X = \begin{pmatrix} i+1 \\ i-1 \\ -1 \end{pmatrix}, \quad Y = \begin{pmatrix} -1 \\ -2 \\ i \end{pmatrix}.$

(c) $X = \begin{pmatrix} -1 \\ 3 \\ i \\ -i \end{pmatrix}, \quad Y = \begin{pmatrix} i+1 \\ -i \\ 2i \\ 1 \end{pmatrix}.$

9. Represent the vector $X = \begin{pmatrix} -1 \\ 0 \\ 2 \end{pmatrix}$ in terms of the orthonormal basis vectors

$$X_1 = \begin{pmatrix} \frac{1}{\sqrt{2}} \\ 0 \\ \frac{1}{\sqrt{2}} \end{pmatrix}, \quad X_2 = \begin{pmatrix} -\frac{\sqrt{2}}{6} \\ \frac{4\sqrt{2}}{6} \\ \frac{\sqrt{2}}{6} \end{pmatrix}, \quad X_3 = \begin{pmatrix} \frac{2}{3} \\ \frac{1}{3} \\ -\frac{2}{3} \end{pmatrix}$$

10. Show that the following vectors form an orthonormal basis:

$$X_1 = \begin{pmatrix} \frac{1}{\sqrt{2}} \\ \frac{i}{\sqrt{2}} \\ 0 \end{pmatrix} \quad X_2 = \begin{pmatrix} \frac{i}{\sqrt{6}} \\ \frac{1}{\sqrt{6}} \\ \frac{2i}{\sqrt{6}} \end{pmatrix}, \quad X_3 = \begin{pmatrix} \frac{-i-1}{\sqrt{6}} \\ \frac{-1+i}{\sqrt{6}} \\ \frac{1+i}{\sqrt{6}} \end{pmatrix}$$

and represent the following in terms of these basis vectors:

$$\text{(a)} \begin{pmatrix} i \\ -1 \\ 6 \end{pmatrix}, \quad \text{(b)} \begin{pmatrix} \frac{-3i}{2} \\ -1 \\ \frac{i}{2} \end{pmatrix}, \quad \text{(c)} \begin{pmatrix} \frac{1}{2} - i \\ \frac{-5}{2}i - 1 \\ i \end{pmatrix}$$

11. Find a method for finding an orthonormal basis from a given basis.

Hint: Let $X_1, X_2, \cdots, X_n$ be the given basis vectors; and $Y_1, Y_2, \cdots Y_n$ the orthonormal basis vectors to be determined. Let

$$Y_1 = X_1/(X_1, X_1)^{1/2},$$

$$Y_2 = \frac{X_2 - (X_2, Y_1)\,Y_1}{(X_2 - (X_2, Y_1)\,Y_1, X_2 - (X_2, Y_1)Y_1)^{1/2}},$$

and generalize these steps. This is known as the Gram-Schmidt process.

12. Consider the following matrices:

$$A = \begin{pmatrix} 1 & 0 & -1 \\ 0 & 3 & 0 \\ 1 & 0 & 0 \end{pmatrix}, \quad B = \begin{pmatrix} 2 & 1 & 0 \\ 0 & 0 & 1 \\ 0 & 1 & 0 \end{pmatrix},$$

$$C = \begin{pmatrix} 1 & 2 \\ 5 & 3 \end{pmatrix}, \quad D = \begin{pmatrix} -1 & 1 \\ 2 & 0 \end{pmatrix}.$$

Find $D - C$, $D + C$, $7A$, $C - D^T$, AB, BA, $(CD)^T$, $(DC)^T$.

13. Construct orthonormal bases from the bases constructed in problem 7.

14. If

$$A = \begin{pmatrix} a & b \\ c & d \end{pmatrix}, \quad B = \begin{pmatrix} 0 & 1 \\ 1 & 0 \end{pmatrix},$$

find $B^{-1}AB$.

15. Find a matrix such that $A^2 = I$.

16. Find matrices corresponding to the following linear transformations. Interpret these transformations geometrically.

(a) $u = 3x, \qquad v = 3y.$

(b) $u = -y, \qquad v = x.$

(c) $u = \dfrac{x - y}{\sqrt{2}}, \qquad v = \dfrac{x + y}{\sqrt{2}}.$

17. Show that multiplication of $n \times n$ diagonal matrices is commutative. Show that if A commutes with all $n \times n$ diagonal matrices, then A must also be diagonal.

18. Show that no real $n \times n$ matrix exists for which $A^2 = -I$, for odd n.

19. Show that $(A^{-1})^T = (A^T)^{-1}$.

20. Find the inverses of the following matrices, if possible. If not, explain why. Check your answers by multiplication.

(a) (5). (b) $\begin{pmatrix} 5 & 1 \\ 6 & 2 \end{pmatrix}$. (c) $\begin{pmatrix} 2 & 1 \\ 3 & 4 \end{pmatrix}$.

(d) $\begin{pmatrix} -1 & 1 \\ 2 & 3 \end{pmatrix}$. (e) $\begin{pmatrix} 4 & 1 & 2 \\ 0 & 1 & 0 \\ 8 & 10 & 4 \end{pmatrix}$. (f) $\begin{pmatrix} 1 & 0 & 1 \\ 0 & 2 & 0 \\ 0 & 0 & 1 \end{pmatrix}$.

21. Find the eigenvalues of the following matrices and find suitable transformation matrices T that diagonalize these matrices:

(a) $\begin{pmatrix} 4 & 1 \\ 0 & 2 \end{pmatrix}$. (b) $\begin{pmatrix} -1 & 1 \\ -5 & 3 \end{pmatrix}$.

(c) $\begin{pmatrix} \frac{2}{3} & \frac{2}{3} & \frac{1}{3} \\ -\frac{2}{3} & \frac{1}{3} & \frac{2}{3} \\ \frac{1}{3} & -\frac{2}{3} & \frac{2}{3} \end{pmatrix}$ (d) $\begin{pmatrix} 4 & 2 & -2 \\ -5 & 3 & 2 \\ -2 & 4 & 1 \end{pmatrix}$.

22. Show that the eigenvalues of a triangular matrix are the diagonal elements.

23. Show that A and A^T have the same eigenvalues.

24. Let T be any nonsingular matrix. Show that A and $T^{-1}AT$ have the same eigenvalues.

25. Show that A is singular if and only if zero is one of its eigenvalues.

26. Let A be a symmetric matrix with distinct and positive eigenvalues. Show that the largest eigenvalue can be characterized as

$$\lambda = \underset{\text{all } X \neq 0}{\text{maximum}} \frac{(X, AX)}{(X, X)}.$$

27. Generalize the process of the preceding problem to characterize all eigenvalues of A by such a maximum property.

28. Show that the following sets of vectors, under the standard operations, form linear vector spaces:

(a) The set of all n-vectors.

(b) The set of all n-vectors such that each of its vectors has the first two components equal to each other.

(c) The set of all n-vectors for which the first three components vanish; $(n \geqslant 3)$.

(d) The set of all n-vectors, such that the sum of the components of each vector vanishes.

What is the dimension of each these spaces?

29. Consider the set of C of all real continuous functions $f(x)$ defined over the interval $0 \leqslant x \leqslant 1$. Show that under the standard operations of addition and multiplication by real scalars, C is a linear vector space.

Show that the space C becomes a normed linear space under norm

$$\| f(x) \| = \max_{0 \leqslant x \leqslant 1} | f(x) |.$$

Show that the following defines an inner product on the space:

$$(f, g) = \int_0^1 f(x)g(x)\, dx.$$

2 Linear Differential Equations

2.1 INTRODUCTION

The most general type of differential equation of order n is of the form

$$F(z, y, y', \cdots, y^{(n)}) = 0. \tag{1}$$

Here z is to be considered as the independent variable, which in general can take on real or complex values; y is the dependent variable. One is interested in finding all functions y that satisfy (1). The most general solution will involve n arbitrary constants; however, if a set of n initial conditions is prescribed in addition to (1), one can determine the arbitrary constants and a unique solution results, provided (1) satisfies some other conditions. More will be said about that later.

If (1) can be written in the form

$$a_0(z)\, y^{(n)} + a_1(z)\, y^{(n-1)}, + \cdots + a_{n-1}(z)\, y' + a_n(z)\, y = f(z), \tag{2}$$

it is called a linear differential equation of order n. Evidently y and its derivatives appear only in linear form. Furthermore the equation is said to be homogenous (inhomogeneous) if the function $f(z)$ on the right side of (2) does (does not) vanish.

2.2 FIRST-ORDER LINEAR DIFFERENTIAL EQUATIONS

For $n = 1$, (2) of Section 2.1 takes the form

$$a_0(z)\, y' + a_1(z)\, y = f(z). \tag{1}$$

It will be assumed that $a_0(z)$, $a_1(z)$, and $f(z)$ are continuous. More general cases must be treated individually. The homogeneous case can be rewritten in the form

$$\frac{d}{dz} \ln y + \frac{a_1(z)}{a_0(z)} = 0,$$

since

$$\frac{d}{dz} \ln y = \frac{1}{y} \frac{dy}{dz},$$

which by an immediate integration takes the form

$$\ln\left(\frac{y}{y_0}\right) + \int_{z_0}^{z} \frac{a_1(t)}{a_0(t)}\, dt = 0.$$

Here the initial condition posed is

$$y(z_0) = y_0.$$

One can solve for y to obtain

$$y = y_0 \exp\left(-\int_{z_0}^{z} \frac{a_1(t)}{a_0(t)}\,dt\right). \tag{2}$$

Here it was implicity assumed that in the z domain under consideration, the function $a_0(z)$ does not vanish.

The special solution

$$y = \exp\left(-\int_{z_0}^{z} \frac{a_1(t)}{a_0(t)}\,dt\right), \tag{3}$$

with the initial value $y(z_0) = 1$, can be considered as a generator of the general solution (2), since (2) is obtained by multiplying (3) by y_0. By a suitable translation of coordinates, the initial point z_0 can be translated into the origin so that without loss of generality we can take $z_0 = 0$.

We now return to (1) and rewrite it in the form

$$y' + \frac{a_1(z)}{a_0(z)}y = \frac{f(z)}{a_0(z)}, \tag{4}$$

with initial value $y(0) = y_0$, and simultaneously we consider the homogeneous equation

$$x' - \frac{a_1(z)}{a_0(z)}x = 0, \tag{5}$$

with initial value $x(0) = 1$, which is known as the adjoint equation of (4). From (3) we find that the solution of (5) is

$$x = \exp\left(\int_{0}^{z} \frac{a_1(t)}{a_0(t)}\,dt\right). \tag{6}$$

To solve (4), we seek a linear differential equation for the product xy. Evidently

$$(xy)' = xy' + x'y = -x\frac{a_1(z)}{a_0(z)}y + x\frac{f(z)}{a_0(z)} + \frac{a_1(z)}{a_0(z)}xy = x\frac{f(z)}{a_0(z)} \tag{7}$$

The advantage of introducing the adjoint equation is that (7) is a very simple differential equation for xy, where the right side is known. By an integration one finds from (7):

$$xy - y_0 = \int_0^z \frac{xf(t)}{a_0(t)}\,dt,$$

so that

$$y = \frac{y_0}{x} + \frac{1}{x}\int_0^z \frac{xf(t)}{a_0(t)}\,dt. \tag{8}$$

One can now replace x from (6) to obtain y explicitly in terms of the functions $a_0(z)$, $a_1(z)$.

EXAMPLE 1. Solve

$$(1 + z^2)\,y' - zy = 0,$$
$$y(0) = 1.$$

Solution

$$\frac{d}{dz}\ln y - \frac{z}{1+z^2} = 0,$$
$$\ln y - \int_0^z \frac{t\,dt}{1+t^2} = 0,$$
$$\ln y - \ln(1+z^2)^{1/2} = 0,$$
$$y = (1+z^2)^{1/2}.$$

EXAMPLE 2. Solve

$$y' - zy = z^3,$$
$$y(0) = 0.$$

Solution. Solve the adjoint equation first:

$$x' + zx = 0,$$
$$x(0) = 1,$$
$$\frac{d}{dz}\ln x + z = 0,$$
$$\ln x + \frac{z^2}{2} = 0,$$
$$x = \exp\left(\frac{-z^2}{2}\right).$$

Then

$$(xy)' = xy' + x'y = x(zy + z^3) - zxy = xz^3 = z^3 \exp\left(\frac{-z^2}{2}\right).$$

By an integration and use of the initial values, one find

$$\begin{aligned} xy &= \int_0^z t^3 \exp\left(\frac{-t^2}{2}\right) dt \\ &= -(2+t^2)\exp\left(\frac{-t^2}{2}\right)\Big|_0^z \\ &= -(2+z^2)\exp\left(\frac{-z^2}{2}\right) + 2, \end{aligned}$$

so that finally

$$y = -2 - z^2 + 2\exp\left(\frac{z^2}{2}\right).$$

One beautiful feature of differential equations is that once a solution has been obtained, it is relatively easy to verify it. For instance, in the preceding example one finds

$$y = -2 - z^2 + 2 \exp\left(\frac{z^2}{2}\right),$$

$$y' = -2z + 2z \exp\left(\frac{z^2}{2}\right),$$

so that

$$y' - zy = z^3,$$

and

$$y(0) = 0,$$

which is the initial equation.

It should be emphasized that the general solution, (8), may sometimes have to be left in a general form. That is, it may not be possible to evaluate the required integral in terms of elementary functions. For example, the solution of the equation

$$y' - zy = z^2,$$
$$y(0) = 0,$$

is given by

$$y = \exp\left(\frac{z^2}{2}\right) \int_0^z t^2 \exp\left(\frac{-t^2}{2}\right) dt.$$

The evaluation of the preceding integral requires the use of higher transcendental functions and therefore will not be carried any further.

An important general question that can now be posed is the following. It has been shown that a solution of (1) exists under the stated hypothesis, but is it the only such solution? In other words, is the solution unique? The answer is in the affirmative and the proof is as follows: Suppose (1) had two solutions, say, y and v. It then follows that the function

$$w = y - v$$

satisfies the differential equation

$$a_0(z)\, w' + a_1(z)\, w = 0,$$
$$w(0) = y(0) - v(0) = 0.$$

This is a homogeneous equation with initial value zero. From (2) one observes that the solution is

$$w = w_0 \exp\left(-\int_0^z \frac{a_1(t)}{a_0(t)}\, dt\right).$$

However, if $w_0 = 0$, it follows that w must vanish identically.

The results of the previous paragraphs can now be summarized in terms of the following theorem.

THEOREM: The differential equation

$$a_0(z)\,y' + a_1(z)\,y = f(z),$$

with the initial value

$$y(0) = y_0,$$

where $a_0(z)$, $a_1(z)$, and $f(z)$ are continuous functions, has a unique solution in every interval including the initial point, provided $a_0(z)$ does not vanish. Furthermore the solution is explicitly furnished by (8).

The continuity requirements stated can on occasion be relaxed considerably; however, the requirement that $a_0(z)$ does not vanish is important, as the following will illustrate. Consider

$$zy' - 2y = 0. \tag{9}$$

From (2) it would follow that the general solution becomes

$$y = y_0 z^2.$$

Evidently it would be impossible to satisfy the initial conditions $y(0) = 1$. In this case no solution would exist.

If one chooses the initial condition $y(1) = 1$, one finds that $y = z^2$, but if the domain of z is to consist of all real z, this solution will not be unique because the function

$$\begin{aligned} v &= 0, & z &\leqslant 0, \\ &= z^2, & z &\geqslant 0, \end{aligned}$$

also satisfies the differential equation and initial condition. Evidently

$$\begin{aligned} v &= y, & z &\geqslant 0, \\ v &\neq y, & z &\leqslant 0. \end{aligned}$$

However, if the domain is restricted to all positive z, in which the coefficient of y' in (9) does not vanish, the only solution with initial value $y(1) = 1$ is

$$y = z^2.$$

The most general class of solutions of (9) with $y(1) = 1$ is given by

$$\begin{aligned} y &= z^2, & z &\geqslant 0, \\ &= cz^2, & z &\leqslant 0, \end{aligned}$$

where c is a completely arbitrary constant. This function is certainly continuous and it also has a continuous derivative, since

$$\begin{aligned} y' &= 2z, \qquad z \geqslant 0, \\ &= 2cz, \qquad z \leqslant 0. \end{aligned}$$

It evidently satisfies the differential equation. However, unless $c = 1$, y is not doubly differentiable. But since we are dealing with a first-order differential equation, we do not require that y possess any derivatives higher than the first.

2.3 SYSTEMS OF DIFFERENTIAL EQUATIONS

In numerous applications one deals not only with a single differential equation in one independent and one dependent variable, but also with a system in one independent and numerous dependent variables. For example,

$$\begin{aligned} F_1(z, y_1, y_1', y_1'', y_2, y_2') &= 0, \\ F_2(z, y_1, y_1', y_2, y_2', y_2'', y_2''') &= 0, \end{aligned} \tag{1}$$

is a system of two equations with two dependent variables. The first equation is of the second order in y_1 and of the first order in y_2, and the second equation is of first order in y_1 and of the third order in y_2. For many purposes it is convenient to rewrite such a system as a system of first-order equations. We shall first assume that we can solve the first of equations (1) for y_1'' and the second for y_2'''. Then

$$\begin{aligned} y_1'' &= G_1(z, y_1, y_1', y_2, y_2'), \\ y_2''' &= G_2(z, y_1, y_1', y_2, y_2', y_2''). \end{aligned} \tag{2}$$

Next we introduce the following five new dependent variables:

$$\begin{aligned} x_1 &= y_1, \\ x_2 &= y_1', \\ x_3 &= y_2, \\ x_4 &= y_2', \\ x_5 &= y_2''. \end{aligned}$$

It follows that (2) can now be rewritten in the form

$$\begin{aligned} x_1' &= x_2, \\ x_2' &= G_1(z, x_1, x_2, x_3, x_4), \\ x_3' &= x_4, \\ x_4' &= x_5, \\ x_5' &= G_2(z, x_1, x_2, x_3, x_4, x_5). \end{aligned} \tag{3}$$

This is a system of five first-order equations. The most general such system is of the form

$$x_i' = G_i(z, x_1, x_2, \cdots, x_n), \qquad i = 1, 2, \cdots, n \tag{4}$$

which is a system of n first-order equations. Here one can achieve an additional notational advantage by introducing two column vectors:

$$X \equiv \begin{pmatrix} x_1 \\ x_2 \\ \vdots \\ x_n \end{pmatrix} \qquad \mathscr{G}(z, X) \equiv \begin{pmatrix} G_1(z, x_1 \cdots x_n) \\ \vdots \\ G_n(z)\, x_1, \cdots, x_n) \end{pmatrix},$$

so that system (4) can be rewritten in the compact form

$$X' = \mathscr{G}(z, X).$$

It is evident that every differential equation of order n can be rewritten as a first-order system of n equations. But the system viewpoint is more general. For example, the system

$$\begin{aligned} x_1' &= a_{11}(z)\, x_1 + a_{12}(z)\, x_2, \\ x_2' &= a_{21}(z)\, x_1 + a_{21}(z)\, x_2, \end{aligned} \tag{5}$$

where the $a_{ij}(z)$ are given functions of z, can be reduced to a second-order differential equation by elimination of x_2. One then obtains

$$\begin{aligned} x_1'' &- \left(a_{11}(z) + a_{22}(z) + \frac{a_{12}'(z)}{a_{12}(z)}\right) x_1' \\ &+ \Bigg[a_{11}(z)\, a_{22}(z) - a_{12}(z)\, a_{21}(z) \\ &- \left(\frac{a_{12}(z)\, a_{11}'(z) - a_{12}'(z)\, a_{11}(z)}{a_{12}(z)}\right)\Bigg]\, x_1 = 0. \end{aligned} \tag{6}$$

It will be shown later that system (5) will, if initial conditions are prescribed, have a unique solution, provided the $a_{ij}(z)$ are continuous functions. But if $a_{12}(z)$ and $a_{11}(z)$ are not differentiable as well, the reduction to (6) will not be possible. This shows that (5) is more general than (6).

2.4 SYSTEMS OF LINEAR DIFFERENTIAL EQUATIONS

The most general linear system of n first-order equations has the form

$$x_i' = \sum_{k=1}^{n} a_{ik}(z)\, x_k + f_i(z), \qquad i = 1, 2, \cdots, n, \tag{1}$$

with initial conditions $x_1(0) = x_{1,0}, x_2(0) = x_{2,0} \cdots x_n(0) = x_{n,0}$, which can be written as a vector system. Let

$$X = \begin{pmatrix} x_1(z) \\ x_2(z) \\ \vdots \\ x_n(z) \end{pmatrix}, \quad A(z) = (a_{ik}(z)), \quad F(z) = \begin{pmatrix} f_1(z) \\ \vdots \\ f_n(z) \end{pmatrix},$$

so that (1) becomes

$$X' = A(z)\,X + F(z)$$

with initial conditions

$$X(0) = \begin{pmatrix} x_1(0) \\ x_2(0) \\ \vdots \\ x_n(0) \end{pmatrix} = \begin{pmatrix} x_{1,0} \\ x_{2,0} \\ \vdots \\ x_{n,0} \end{pmatrix} = X_0 \tag{2}$$

Such a system will be called homogeneous (inhomogeneous) if the function $F(z)$ does (does not) vanish. Here one can show, in analogy to the theorem of Section 2.2, that if suitable conditions are prescribed on $A(z)$ and $F(z)$ and an initial vector for X is prescribed, $X(0) = X_0$, then the system (2) will have a unique solution.

2.5 HOMOGENEOUS SYSTEMS OF LINEAR DIFFERENTIAL EQUATIONS WITH CONSTANT COEFFICIENTS

We now turn our attention to the equation

$$X' = AX \tag{1}$$

where A is a constant $n \times n$ matrix. Such a system has precisely n linearly independent solution vectors.

We shall for the moment use the uniqueness theorem, which will be proved later. With it one can show that there can be no more than n linearly independent solutions. Suppose $X_1, X_2, \cdots, X_m$ are m linearly independent solutions where $m > n$. Then the vector

$$X = \sum_{i=1}^{m} c_i X_i$$

is a solution of (1). We now choose the scalars c_i so that

$$\sum_{i=1}^{m} c_i X_i(0) = 0.$$

This system must have nonvanishing solutions, since it represents a system of n equations in $m > n$ unknowns. But then

$$X(0) = 0,$$

so that $X \equiv 0$ by the uniqueness theorem, showing the m X_i are linearly dependent.

We shall now demonstrate that precisely n linearly independent vectors can be found. First we shall consider a relatively simple case, namely, the case where all eigenvalues are distinct; in this case the matrix A can be diagonalized. Then a matrix T can be found such that

$$T^{-1}AT = D$$

where

$$D = \begin{pmatrix} \lambda_1 & & & & 0 \\ & \lambda_2 & & & \\ & & \cdot & & \\ & & & \cdot & \\ 0 & & & & \lambda_n \end{pmatrix}$$

and the λ_i represent all eigenvalues of A. Then, if a new set of dependent variables is introduced, defined by

$$Y = T^{-1}X, \tag{2}$$

one finds by substitution in (1) that

$$Y' = T^{-1}X' = T^{-1}AX = T^{-1}ATY = DY. \tag{3}$$

The preceding system when written out in scalar notation has the simple form

$$y_i' = \lambda_i y_i, \qquad i = 1, 2, \cdots, n, \tag{4}$$

where each equation contains only one dependent variable. These are first-order equations, which were treated in Section 2.2, and one can see immediately that

$$y_i = c_i e^{\lambda_i z}$$

where the c_i are n arbitrary scalars. One can now construct n linearly independent solutions for the system (3). For example, one such set of solutions is given by

$$Y_1 = \begin{pmatrix} e^{\lambda_1 z} \\ 0 \\ 0 \\ \vdots \\ 0 \\ 0 \end{pmatrix}, \quad Y_2 = \begin{pmatrix} 0 \\ e^{\lambda_2 z} \\ 0 \\ \vdots \\ 0 \\ 0 \end{pmatrix}, \cdots, \quad Y_n = \begin{pmatrix} 0 \\ 0 \\ 0 \\ \vdots \\ 0 \\ e^{\lambda_n z} \end{pmatrix} \tag{5}$$

To verify that these are truly linearly independent, one merely observes that the determinant formed from these vectors,

$$\begin{vmatrix} e^{\lambda_1 z} & 0 & & 0 \\ 0 & e^{\lambda_2 z} & \cdots & 0 \\ 0 & 0 & & 0 \\ \vdots & \vdots & & \\ 0 & 0 & & 0 \\ 0 & 0 & & e^{\lambda_n z} \end{vmatrix} = e^{(\lambda_1+\lambda_2+\cdots+\lambda_n)z},$$

does not vanish. It is now evident that (1) has the n linearly independent vector solutions

$$TY_1, TY_2, \cdots, TY_n.$$

Such a set of vectors is known as a fundamental set of solutions of (1). From these one can construct n other solutions of (1):

$$X_1, X_2, \cdots, X_n$$

with the particular initial vectors

$$X_i(0) = \begin{pmatrix} \delta_{1i} \\ \delta_{2i} \\ \vdots \\ \delta_{ni} \end{pmatrix}, \qquad i = 1, 2, \cdots n,$$

where δ_{ij} is the Kronecker δ.

The matrix Φ, whose ith column is equal to X_i satisfies the differential equation

$$\Phi' = A\Phi \tag{6}$$

with the initial value

$$\Phi(0) = I.$$

The equation (6) is similar to (1) except that Φ is a matrix whereas X is a column vector. But every column of Φ is a solution of (1) .

It is now very easy to find a solution of (1) with the initial vector

$$X(0) = X_0.$$

This solution is explicitly furnished by

$$X = \Phi X_0. \tag{7}$$

To verify that this is the required solution, we observe first that

$$X(0) = \Phi(0)\, X_0 = IX_0 = X_0$$

and that

$$X' = \Phi' X_0 = A\Phi X_0 = AX,$$

thus satisfying the differential equation and initial condition.

The method followed in the preceding paragraphs is satisfactory as long as A has distinct eigenvalues, but may fail in more general situations. But it shows very clearly that the system (1) will have exponential solutions. We shall therefore attempt to find a solution vector in the form

$$X = e^{\lambda t} C$$

where λ is still unknown and C is an unknown constant vector. Insertion of this trial solution in (1) leads to

$$\lambda e^{\lambda t} C = e^{\lambda t} AC,$$

from which one immediately deduces the equation

$$(A - \lambda I)\, C = 0.$$

Such a homogeneous system will have a nonvanishing solution C if and only if

$$| A - \lambda I | = 0.$$

Here again we see that if A has distinct eigenvalues, one can find n eigenvalues λ_i and corresponding eigenvectors C_i from which the general solution of (1)

$$X = \sum_{i=1}^{n} e^{\lambda_i t} C_i$$

is constructed.

EXAMPLE 1. Solve

$$y''' + 6y'' + 11y' + 6y = 0,$$

where

$$y(0) = 1, \qquad y'(0) = y''(0) = 0.$$

Solution. Let

$$x_1 = y,$$
$$x_2 = y',$$
$$x_3 = y'',$$

so that

$$\begin{aligned} x_1' &= x_2, \\ x_2' &= x_3, \\ x_3' &= y''' = -6y - 11y' - 6y'' \\ &= -6x_1 - 11x_2 - 6x_3, \end{aligned}$$

which can be written as

$$X' = \begin{pmatrix} 0 & 1 & 0 \\ 0 & 0 & 1 \\ -6 & -11 & -6 \end{pmatrix} X,$$

where

$$X(0) = \begin{pmatrix} 1 \\ 0 \\ 0 \end{pmatrix}.$$

To find solutions in the form

$$X = e^{\lambda t} C,$$

we see that

$$\lambda e^{\lambda t} C = e^{\lambda t} \begin{pmatrix} 0 & 1 & 0 \\ 0 & 0 & 1 \\ -6 & -11 & -6 \end{pmatrix} C,$$

which leads to the equation

$$\begin{pmatrix} -\lambda & 1 & 0 \\ 0 & -\lambda & 1 \\ -6 & -11 & -6-\lambda \end{pmatrix} C = 0.$$

Hence λ must satisfy the equation

$$\begin{vmatrix} -\lambda & 1 & 0 \\ 0 & -\lambda & 1 \\ -6 & -11 & -6-\lambda \end{vmatrix} = -\lambda^3 - 6\lambda^2 - 11\lambda - 6$$
$$= -(\lambda + 1)(\lambda + 2)(\lambda + 3) = 0,$$

so that there are three distinct eigenvalues:

$$\lambda_1 = -1, \qquad \lambda_2 = -2, \qquad \lambda_3 = -3.$$

One can now determine the corresponding eigenvectors

$$C_1 = \begin{pmatrix} a \\ -a \\ a \end{pmatrix}, \qquad C_2 = \begin{pmatrix} b \\ -2b \\ 4b \end{pmatrix}, \qquad C_3 = \begin{pmatrix} c \\ -3c \\ 9c \end{pmatrix},$$

where a, b, and c are to be determined from the initial condition

$$X(0) = \begin{pmatrix} 1 \\ 0 \\ 0 \end{pmatrix}.$$

This leads to the equations

$$C_1 + C_2 + C_3 = \begin{pmatrix} 1 \\ 0 \\ 0 \end{pmatrix},$$

or

$$\begin{pmatrix} a + b + c \\ -a - 2b - 3c \\ a + 4b + 9c \end{pmatrix} = \begin{pmatrix} 1 \\ 0 \\ 0 \end{pmatrix}.$$

One then finds that

$$a = 3, \qquad b = -3, \qquad c = 1,$$

so that the final solution becomes

$$\begin{aligned} X &= e^{-t}\begin{pmatrix} 3 \\ -3 \\ 3 \end{pmatrix} + e^{-2t}\begin{pmatrix} -3 \\ 6 \\ -12 \end{pmatrix} + e^{-3t}\begin{pmatrix} 1 \\ -3 \\ 9 \end{pmatrix} \\ &= \begin{pmatrix} 3e^{-t} - 3e^{-2t} + e^{-3t} \\ -3e^{-t} + 6e^{-2t} - 3e^{-3t} \\ 3e^{-t} - 12e^{-2t} + 9e^{-3t} \end{pmatrix}. \end{aligned}$$

Returning to the original third-order equation, we have $y = 3e^{-t} - 3e^{-2t} + e^{-3t}$.

So far the general case where multiple eigenvalues might occur has not been considered. To treat this case, we shall first show that a general $n \times n$ system can always be reduced to an $(n - 1) \times (n - 1)$ system, once one solution has been obtained. We now return to the system

$$X' = AX$$

and assume that a particular solution

$$X_1 = \begin{pmatrix} x_{11} \\ x_{12} \\ \vdots \\ x_{1n} \end{pmatrix},$$

which does not vanish identically, has been found. Next we define the matrix

$$\Phi = \begin{vmatrix} 1 & 0 & \cdots & 0 & x_{11} \\ 0 & 1 & & 0 & x_{12} \\ 0 & 0 & & 0 & x_{13} \\ \vdots & \vdots & & \vdots & \vdots \\ 0 & 0 & & 1 & x_{1n-1} \\ 0 & 0 & & 0 & x_{1n} \end{vmatrix} \tag{8}$$

and assume that x_{1n} does not vanish, in which case the matrix is nonsingular. If x_{1n} were to vanish, we could assume that for some k, x_{1k} does not vanish; otherwise, X_1 would vanish, which is not true by assumption. In this case we should simply use a matrix similar to Φ; that is, one that is similar to an identity matrix, except that we should replace the kth column by X_1. This, too, would be nonsingular, and then we should proceed exactly as we shall with the matrix Φ in (8).

Next, a new set of dependent variables Y will be introduced, which are related to X by

$$X = \Phi Y.$$

Insertion of this substitution in (1) leads to

$$\Phi' Y + \Phi Y' = A\Phi Y,$$

which can be solved for Y' to yield

$$Y' = \Phi^{-1}(A\Phi - \Phi')\, Y \equiv BY \tag{9}$$

The B matrix is defined by (9). One can easily see that a particular solution of (9) is given by

$$Y_1 = \begin{pmatrix} 0 \\ 0 \\ \vdots \\ 0 \\ 1 \end{pmatrix},$$

since

$$\Phi Y_1 = X_1,$$

the particular solution that the system (1) has. From this solution, Y_1, one can get some insight into the structure of the matrix B in (9). Insertion of Y_1 into (9) leads to

$$\begin{pmatrix} 0 \\ 0 \\ 0 \\ \vdots \\ 0 \end{pmatrix} = B \begin{pmatrix} 0 \\ 0 \\ \vdots \\ 1 \end{pmatrix} = \begin{pmatrix} b_{1n} \\ b_{2n} \\ \vdots \\ b_{nn} \end{pmatrix}.$$

Thus we see that

$$b_{in} = 0, \qquad i = 1, 2, \cdots, n.$$

When (9) is now written out as a scalar system, one has

$$y_i' = \sum_{j=1}^{n-1} b_{ij} y_j, \qquad i = 1, 2, \cdots, n-1,$$

$$y_n' = \sum_{j=1}^{n-1} b_{nj} y_j. \tag{10}$$

This shows that the first $(n-1)$ components of Y satisfy an $(n-1)\times(n-1)$ system, and once that is solved, y_n can be found from the second of the equations in (10).

One can show that if one knows k particular solutions, then by a similar process one can reduce an $n \times n$ system to an $(n-k) \times (n-k)$ system. Suppose that

$$X_i = \begin{pmatrix} x_{1i} \\ x_{2i} \\ \vdots \\ x_{ni} \end{pmatrix}, \qquad i = 1, 2, \cdots k$$

are k known solutions of (1). A set of new independent variables, Y, defined by

$$X = \Phi Y,$$

where

$$\begin{pmatrix} 1 & 0 & \cdots & x_{11} & \cdots & x_{1k} \\ 0 & 1 & \cdots & x_{21} & & x_{2k} \\ 0 & 0 & & & & \\ \vdots & \vdots & & \vdots & & \vdots \\ 0 & 0 & & x_{n1} & & x_{nk} \end{pmatrix}$$

can be introduced. This effectively reduces the system to an $(n-k) \times (n-k)$ system.

EXAMPLE 2. Find the general solution of

$$y''' + 4y'' + 5y' + 2y = 0.$$

Solution. Let

$$x_1 = y,$$
$$x_2 = y',$$
$$x_3 = y'',$$

so that

$$X' = \begin{pmatrix} 0 & 1 & 0 \\ 0 & 0 & 1 \\ -2 & -5 & -4 \end{pmatrix} X.$$

If we use the trial solution

$$X = e^{\lambda t}C,$$

we find that λ must satisfy the equation

$$\begin{vmatrix} -\lambda & 1 & 0 \\ 0 & -\lambda & 1 \\ -2 & -5 & -4-\lambda \end{vmatrix} = -\lambda^3 - 4\lambda^2 - 5\lambda - 2$$
$$= -(\lambda + 1)^2 (\lambda + 2) = 0,$$

so that we have only two distinct eigenvalues,

$$\lambda_1 = -1, \qquad \lambda_2 = -2,$$

and the corresponding eigenvectors are

$$C_1 = \begin{pmatrix} a \\ -a \\ a \end{pmatrix}, \qquad C_2 = \begin{pmatrix} b \\ -2b \\ 4b \end{pmatrix},$$

where a and b are arbitrary. To find a third linearly independent solution, we construct Φ:

$$\Phi = \begin{pmatrix} 1 & 0 & ae^{-t} \\ 0 & 1 & -ae^{-t} \\ 0 & 0 & ae^{-t} \end{pmatrix}$$

with

$$\Phi^{-1} = \begin{pmatrix} 1 & 0 & -1 \\ 0 & 1 & 1 \\ 0 & 0 & (1/a)\, e^{t} \end{pmatrix}.$$

Let

$$X = \Phi Y$$

so that

$$Y' = \Phi^{-1}(A\Phi - \Phi')\, Y = \begin{pmatrix} 2 & 6 & 0 \\ -2 & -5 & 0 \\ -2e^t/a & -5e^t/a & 0 \end{pmatrix} Y.$$

In scalar notation we have

$$\begin{aligned} y_1' &= 2y_1 + 6y_2, \\ y_2' &= -2y_1 - 5y_2, \\ y_3' &= \left(\frac{-2}{a}\right) e^t y_1 - \left(\frac{5}{a}\right) e^t y_2, \end{aligned} \tag{11}$$

so that once y_1 and y_2 have been obtained from the first two equations, y_3 can be found by an integration. One then finds that

$$\begin{aligned} y_1 &= -2ce^{-t}, \\ y_2 &= ce^{-t}, \\ y_3 &= \left(\frac{-c}{a}\right) t. \end{aligned} \tag{12}$$

To find X_3, we now compute

$$X_3 = \Phi \begin{pmatrix} -2ce^{-t} \\ ce^{-t} \\ (-c/a)\,t \end{pmatrix} = e^{-t} \begin{pmatrix} -2c \\ c \\ -0 \end{pmatrix} + te^{-t} \begin{pmatrix} -c \\ c \\ -c \end{pmatrix},$$

where c is arbitrary. It follows that the general solution can be put into the form

$$X = e^{-t} \begin{pmatrix} a - 2c \\ -a + c \\ a \end{pmatrix} + te^{-t} \begin{pmatrix} -c \\ c \\ -c \end{pmatrix} + e^{-2t} \begin{pmatrix} b \\ -2b \\ 4b \end{pmatrix}. \tag{13}$$

One can now easily verify that three linearly independent solutions have been found.

Since y_1 and y_2 are found to be multiples of e^{-t}, the inhomogeneous term in the third equation of (11) turns out to be a constant, so that an integration leads to a linear term in t in the third equation of (12). As a consequence of λ_1 being a double eigenvalue, one finds therefore that the general solution (13) does not involve only exponentials but also terms of the type $te^{\lambda t}$. To show that this is a general conclusion, we return to system (9) and examine the B matrix. Since Φ is a function of t, B will no longer be a constant matrix, but an explicit computation of B shows that only the elements of the nth row will depend explicitly on t. The first $n - 1$ rows are constant.

A computation shows that if the particular solution used is of the form

$$X_1 = e^{\lambda_1 t} C_1,$$

then the elements of the last row of B have the form

$$\begin{aligned} b_{nj} &= e^{-\lambda_1 t} \beta_{nj}, \qquad j = 1, 2, \cdots, n - 1, \\ b_{nn} &= 0 \end{aligned}$$

where the β_{nj} are constant. These conclusions are clearly illustrated in Example 2.

The reduced $(n - 1) \times (n - 1)$ system will also have exponential solutions in which case y_n will also be exponential unless the eigenvalue of the reduced system that is used is also λ_1. In this case the equation for y_n' becomes

$$y_n' = \sum_{j=1}^{n-1} b_{nj} y_j = \sum_{j=1}^{n-1} e^{-\lambda_1 t} \beta_{nj} e^{\lambda_1 t} c_j = \sum_{j=1}^{n-1} \beta_{nj} c_j.$$

In this and only in this case will y_n contain linear t terms, and then the general solution will have to contain terms of the form $te^{\lambda_1 t}$. Evidently this can occur only when λ_1 is a double eigenvalue. Otherwise, as shown previously, only solutions of the form $e^{\lambda t}$ can occur where λ takes on all n distinct eigenvalues and there can be at most n linearly independent solutions.

In case λ_1 is a triple eigenvalue, X_1 will be exponential, and a second solution X_2 will contain terms of the form $te^{\lambda_1 t}$. The reduced $(n-1) \times (n-1)$ system, obtained by the use of X_1 will contain an exponential solution Y_2 of the form $e^{\lambda_1 t}$, which leads to an X_2 of the form $te^{\lambda_1 t}$. But if the $(n-1) \times (n-1)$ system is reduced to an $(n-2) \times (n-2)$ system by the use of Y_2, then a Y_3 will be found of the form $te^{\lambda_1 t}$, leading to an X_3 of the form $t^2 e^{\lambda_1 t}$. From these considerations we can formulate the following theorem.

THEOREM: The system

$$X' = AX,$$

where

$$| A - \lambda I | = 0,$$

has roots $\lambda_1, \lambda_2, \cdots \lambda_k$ of multiplicities $m_1, m_2, \cdots m_k$, respectively, has a general solution of the form

$$X = \sum_{j=1}^{k} \sum_{i=1}^{m_j} t^{i-1} e^{\lambda_j t} C_{ij}.$$

Evidently $m_1 + m_2 + \cdots + m_k = n$ and the sum consists of n linearly independent terms.

EXAMPLE 3. Solve

$$X' = \begin{pmatrix} 0 & 1 & 0 \\ 0 & 0 & 1 \\ -1 & -3 & -3 \end{pmatrix} X.$$

Solution. $| A - \lambda I | = -(\lambda + 1)^3$ so that there is a triple root at $\lambda = -1$. According to the preceding theorem, the general solution must have the form

$$X = e^{-t} C_1 + te^{-t} C_2 + t^2 e^{-t} C_3.$$

After insertion in the equation, one obtains

$$e^{-t}(-C_1 + C_2) + te^{-t}(-C_2 + 2C_3) + t^2 e^{-t}(-C_3) = e^{-t} AC_1 + te^{-t} AC_2 + t^2 e^{-t} AC_3.$$

If coefficients of corresponding terms are now equated, one obtains

$$\begin{aligned} AC_1 &= -C_1 + C_2, \\ AC_2 &= -C_2 + 2C_3, \\ AC_3 &= -C_3. \end{aligned}$$

The third of these equations is homogeneous and has a vanishing determinant. We find that

$$C_3 = \begin{pmatrix} a \\ -a \\ a \end{pmatrix}.$$

The second equation now becomes

$$\begin{pmatrix} 1 & 1 & 0 \\ 0 & 1 & 1 \\ -1 & -3 & -2 \end{pmatrix} C_2 = \begin{pmatrix} 2a \\ -2a \\ 2a \end{pmatrix}.$$

This system has determinant zero, but is compatible and therefore has a solution, which is not unique; it has another arbitrary element so that

$$C_2 = \begin{pmatrix} b \\ 2a - b \\ -4a + b \end{pmatrix}.$$

From the first equation one now finds in a similar manner that

$$C_1 = \begin{pmatrix} c \\ b - c \\ 2a - 2b + c \end{pmatrix}.$$

Then

$$X = e^{-t} \begin{pmatrix} c \\ b - c \\ 2a - 2b + c \end{pmatrix} + te^{-t} \begin{pmatrix} b \\ 2a - b \\ -4a + b \end{pmatrix} + t^2 e^{-t} \begin{pmatrix} a \\ -a \\ a \end{pmatrix},$$

where the three arbitrary constants a, b, c could be determined from initial conditions.

In a purely formal sense, (1), namely,

$$X' = AX$$

with initial vector

$$X(0) = X_0$$

has the solution

$$X = e^{tA} X_0, \tag{14}$$

in complete analogy to the case of a simple first-order equation. The function e^{tA} must be interpreted as a Taylor series:

$$e^{tA} = \sum_{n=0}^{\infty} \frac{t^n A^n}{n!}.$$

That this series converges follows from the comparison test with an ordinary exponential series

$$\left\| \frac{t^n A^n}{n!} \right\| \leq \frac{|t|^n}{n!} \| A \|^n,$$

and it is known that

$$\sum_{n=0}^{\infty} \frac{u^n}{n!}$$

converges for all finite values of u, so that we can conclude from

$$\left\| \sum_{k=0}^{n} \frac{t^k A^k}{k!} \right\| \leq \sum_{k=0}^{n} \frac{|t|^k}{k!} \| A \|^k$$

that the series will converge for every constant matrix A and all finite values of t. One can also show that

$$\frac{d}{dt} e^{tA} = \frac{d}{dt} \sum_{0}^{\infty} \frac{t^k A^k}{k!} = \sum_{0}^{\infty} \frac{k t^{k-1} A^k}{k!} = A \sum_{k=1}^{\infty} \frac{t^{k-1} A^{k-1}}{(k-1)!} = A e^{tA},$$

so that (14) satisfies (1) and also has the initial value

$$X(0) = e^{0 \cdot A} X_0 = I X_0 = X_0.$$

The formal solution (14) is in general not useful unless one can sum the infinite series in some convenient way. For example, when A has distinct eigenvalues, one can find a matrix T such that

$$T^{-1} A T = D,$$

where D is diagonal. Then one can also write A in the form

$$A = T D T^{-1},$$

from which one deduces

$$A^2 = (T D T^{-1})(T D T^{-1}) = T D^2 T^{-1}$$

by rearranging the terms in the product by the associative rule of multiplication.

One can now show very easily (by mathematical induction, for instance) that in general

$$A^n = TD^nT^{-1}.$$

Then one finds that

$$e^{tA} = \sum_{k=0}^{\infty} \frac{t^k A^k}{k!} = \sum_{k=0}^{\infty} \frac{t^k TD^k T^{-1}}{k!} = T \sum_{k=0}^{\infty} \frac{t^k D^k}{k!} T^{-1} = Te^{tD}T^{-1},$$

and solution (14) can be put into the form

$$X = Te^{tD}T^{-1}X_0. \tag{15}$$

What makes this last expression so useful is that the required infinite series can be summed easily and explicitly, as follows:

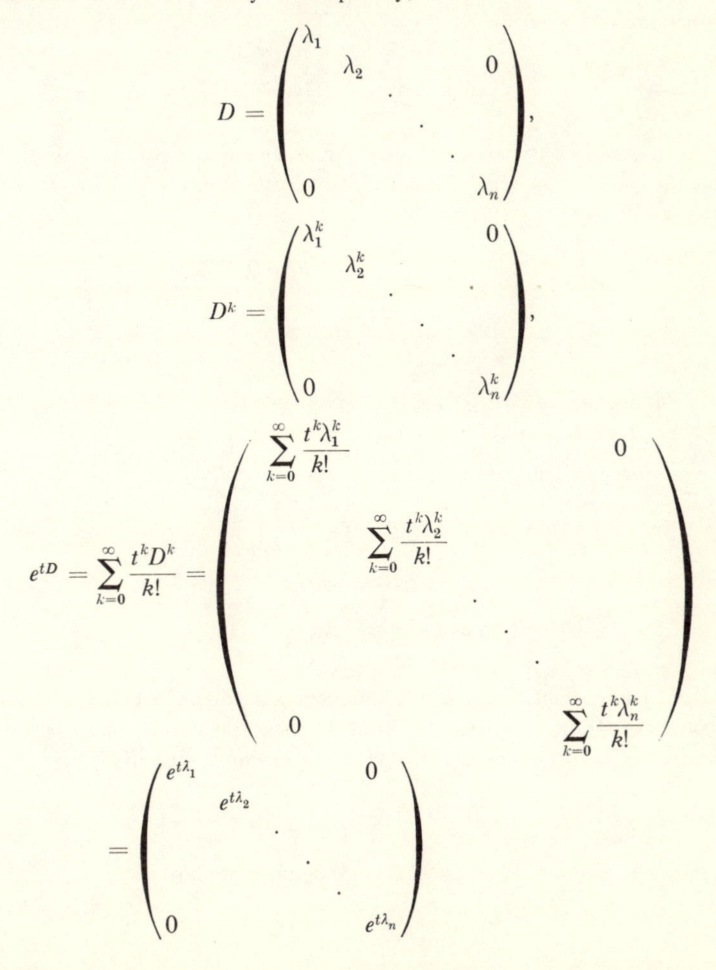

$$D = \begin{pmatrix} \lambda_1 & & & & 0 \\ & \lambda_2 & & & \\ & & \cdot & & \\ & & & \cdot & \\ & & & & \cdot \\ 0 & & & & \lambda_n \end{pmatrix},$$

$$D^k = \begin{pmatrix} \lambda_1^k & & & & 0 \\ & \lambda_2^k & & & \\ & & \cdot & & \\ & & & \cdot & \\ & & & & \cdot \\ 0 & & & & \lambda_n^k \end{pmatrix},$$

$$e^{tD} = \sum_{k=0}^{\infty} \frac{t^k D^k}{k!} = \begin{pmatrix} \sum_{k=0}^{\infty} \frac{t^k \lambda_1^k}{k!} & & & & 0 \\ & \sum_{k=0}^{\infty} \frac{t^k \lambda_2^k}{k!} & & & \\ & & \cdot & & \\ & & & \cdot & \\ & & & & \cdot \\ 0 & & & & \sum_{k=0}^{\infty} \frac{t^k \lambda_n^k}{k!} \end{pmatrix}$$

$$= \begin{pmatrix} e^{t\lambda_1} & & & & 0 \\ & e^{t\lambda_2} & & & \\ & & \cdot & & \\ & & & \cdot & \\ & & & & \cdot \\ 0 & & & & e^{t\lambda_n} \end{pmatrix}$$

so that (15) assumes the form

$$X = T \begin{pmatrix} e^{t\lambda_1} & & & & 0 \\ & e^{t\lambda_2} & & & \\ & & \cdot & & \\ & & & \cdot & \\ & & & & \cdot \\ 0 & & & & e^{t\lambda_n} \end{pmatrix} T^{-1}X_0. \tag{16}$$

2.6 THE LAPLACE TRANSFORM

By using the Laplace transform technique, differential equations with constant coefficients can be converted into algebraic equations. By definition the function $y(s)$, given by

$$y(s) = \int_0^\infty e^{-sz}\, x(z)\, dz, \tag{1}$$

is called the Laplace transform of $x(z)$. We assume that the functions $x(z)$ are so chosen that (1) converges for sufficiently large values of s. One can easily derive the following transforms:

$$\begin{aligned} \int_0^\infty e^{-sz} z^n\, dz &= \frac{n!}{s^{n+1}}, \qquad n = 0, 1, 2, \cdots, \\ \int_0^\infty e^{-sz} e^{\lambda z}\, dz &= \frac{1}{s - \lambda}, \qquad s > \lambda, \\ \int_0^\infty e^{-sz} e^{\lambda z} z^n\, dz &= \frac{n!}{(s - \lambda)^{n+1}}. \end{aligned} \tag{2}$$

By integrating by parts, one finds that

$$\begin{aligned} \int_0^\infty e^{-sz} x'(z)\, dz &= e^{-sz} x(z) \Big|_0^\infty + s \int_0^\infty e^{-sz} x(z)\, dz \\ &= s y(s) - x(0), \\ \int_0^\infty e^{-sz} x''(z)\, dz &= s^2 y(s) - s x(0) - x'(0). \end{aligned} \tag{3}$$

There is an important theorem, the uniqueness theorem, which states that if two continuous functions have the same Laplace transforms, then they must be identical. From this one can conclude, for example, that if

$$\int_0^\infty e^{-sz} x(z)\, dz = \frac{1}{(s - \lambda)^2},$$

then the only continuous function $x(z)$ satisfying the above is

$$x(z) = z e^{\lambda z}.$$

EXAMPLE 1. Solve

$$x' - 5x = 1,$$
$$x(0) = 2.$$

Solution. We apply the Laplace transform to the equation and obtain

$$\int_0^\infty e^{-sz}(x'(z) - 5x(z))\, dz = \int_0^\infty e^{-sz}\, 1\, dz.$$

After evaluation of the integrals the following algebraic equation for $y(s)$ is found:

$$sy(s) - 2 - 5y(s) = \frac{1}{s},$$

$$y(s) = \frac{1/s + 2}{s - 5} = \frac{1 + 2s}{s(s - 5)}.$$

Thus the Laplace transform of the solution has been found, and the corresponding $x(z)$ must be found. First the rational function on the right will be expanded into partial fractions so that

$$y(s) = \frac{-1/5}{s} + \frac{11/5}{s - 5}.$$

From (2) one sees that the only continuous $x(z)$ with the above transform is

$$x(z) = -\left(\frac{1}{5}\right) + \left(\frac{11}{5}\right) e^{5z},$$

which is the solution of the given differential equation.

This method can be extended to systems of differential equations with constant coefficients. We consider the homogeneous system

$$\begin{aligned} X' &= AX, \\ X(0) &= X_0. \end{aligned} \tag{4}$$

We now let

$$Y(s) = \begin{pmatrix} y_1(s) \\ \vdots \\ y_n(s) \end{pmatrix} = \int_0^\infty e^{-sz}\, X(z)\, dz = \begin{pmatrix} \int_0^\infty e^{-sz} \; x_1(z)\, dz \\ \vdots \\ \int_0^\infty e^{-sz} \; x_n(z)\, dz \end{pmatrix},$$

and as in (3),

$$sY(s) - X_0 = \int_0^\infty e^{-sz} X'(z)\, dz.$$

An application of the Laplace transform to (4) leads to

$$\int_0^\infty e^{-sz} X'(z)\, dz = \int_0^\infty e^{-sz} AX(z)\, dz = A \int_0^\infty e^{-sz} X(z)\, dz,$$

$$sY(s) - X_0 = AY(s),$$

$$(sI - A)\, Y(s) = X_0. \tag{5}$$

Hence (5) is a system of algebraic equations for $Y(s)$. If $B(s)$ is now defined by

$$B(s)\,(sI - A) = (sI - A)\, B(s) = I, \tag{6}$$

that is, $B(s)$ is the inverse of $sI - A$, and of course is a function of s, we find

$$Y(s) = B(s)\, X_0. \tag{7}$$

Thus $Y(s)$ has been found, and all that is left to do is to find an $X(z)$ whose transform is $Y(s)$. To do this, we have to investigate $B(s)$ in greater detail. Now

$$\Delta(s) = |\, sI - A\, |$$

is the characteristic polynomial of A. Its zeros are the eigenvalues of A. Also we have

$$|\, B(s)\,(sI - A)\, | = |\, B(s)\, |\, \Delta(s) = 1$$

by the rule regarding determinants of products of matrices, so that

$$|\, B(s)\, | = \frac{1}{\Delta(s)}\,.$$

From the general method of constructing inverses, one can see that every component of $B(s)$ must be a rational function of s. Furthermore, $sI - A$ must have an inverse for every value of s, with the exception of eigenvalues of A. In other words, the (i, j) component of $B(s)$ has the form

$$\frac{p_{ij}(s)}{\Delta(s)}\,,$$

where $p_{ij}(s)$ is a polynomial, since the denominator can vanish only if s is a zero of $\Delta(s)$. Now (6) can be rewritten in the form

$$\left(I - \frac{1}{s} A\right) sB(s) = I,$$

and it follows that in the limit as s approaches infinity,

$$\lim_{s \to \infty} sB(s) = I.$$

Therefore

$$\lim_{s\to\infty} B(s) = 0,$$

and it follows that the degree of $p_{ij}(s)$ must be less than n.

First we shall examine the case where $\Delta(s)$ has only distinct eigenvalues, say, $\lambda_1, \lambda_2, \cdots \lambda_n$. Then, since each component of $B(s)$ can be expanded into partial fractions, we can expand $B(s)$ into

$$B(s) = \sum_{i=1}^{n} \frac{B_i}{s - \lambda_i},$$

where the B_i are constant matrices. Returning to (7) we find

$$Y(s) = \sum_{i=1}^{n} \frac{B_i X_0}{s - \lambda_i}$$

so that by using (2), we find

$$X(z) = \sum_{i=1}^{n} e^{\lambda_i z} B_i X_0. \tag{8}$$

This is, of course, in full agreement with the results of the preceding section.

In the more general case where multiple eigenvalues are allowed, we have

$$\Delta(s) = (s - \lambda_1)^{m_1}(s - \lambda_2)^{m_2} \cdots (s - \lambda_k)^{m_k},$$

and the partial fraction decomposition takes the form

$$B(s) = \sum_{i=1}^{k} \sum_{j=1}^{m_i} \frac{B_{ij}}{(s - \lambda_i)^j}, \tag{9}$$

where the B_{ij} are constant matrices. Then we have from (7),

$$Y(s) = \sum_{i=1}^{k} \sum_{j=1}^{m_i} \frac{B_{ij} X_0}{(s - \lambda_i)^j},$$

so that, from (2),

$$X(z) = \sum_{i=1}^{k} \sum_{j=1}^{m_i} \frac{z^{j-1} e^{\lambda_i z} B_{ij} X_0}{(j-1)!}. \tag{10}$$

This result was found in the preceding section, but once the decomposition (9) is found, (10) follows immediately.

One can also obtain more information regarding the structure of the matrix A. From (6) and (9) we have

$$(sI - A)\sum_{i=1}^{k}\sum_{j=1}^{m_i}\frac{B_{ij}}{(s-\lambda_i)^j} = I, \tag{11}$$

and in particular if all λ_i are simple eigenvalues,

$$(sI - A)\sum_{i=1}^{n}\frac{B_i}{s-\lambda_i} = I.$$

If the above is multiplied by $s - \lambda_l$, and then s is allowed to approach λ_l, we find

$$(\lambda_l I - A)\, B_l = 0. \tag{12}$$

If X_l is a vector formed by taking one of the nonvanishing columns of B_l, we must have

$$(\lambda_l I - A)\, X_l = 0.$$

Then, as shown in Section 1.6, one can construct a nonsingular matrix T by taking as its columns $X_1, X_2, \cdots, X_n$, and then

$$T^{-1}AT = \begin{pmatrix} \lambda_1 & & & 0 \\ & \lambda_2 & & \\ & & \ddots & \\ 0 & & & \lambda_n \end{pmatrix},$$

a diagonal matrix.

To consider the more general case, we return to (11) and write it in the form

$$[(s-\lambda_l)\,I + (\lambda_l I - A)]\left[\frac{B_{lm_l}}{(s-\lambda_l)^{m_l}} + \frac{B_{lm_l-1}}{(s-\lambda_l)^{m_l-1}} + \cdots + \frac{B_{l1}}{s-\lambda_l} + Q\right] = I,$$

where Q contains all terms when $i \neq l$. If the multiplication indicated is carried out, one finds

$$\frac{(\lambda_l I - A)\, B_{lm_l}}{(s-\lambda_l)^{m_l}} + \frac{B_{lm_l} + (\lambda_l I - A)\, B_{lm_l-1}}{(s-\lambda_l)^{m_l-1}} + \cdots + \frac{B_{l2} + (\lambda_l I - A)\, B_{l1}}{s-\lambda_l}$$
$$+ B_{l1} + (s-\lambda_l)\,Q + (\lambda_l I - A)\,Q = I.$$

Since the right side has no singularities, all singular terms on the left must vanish. Then

$$\begin{aligned} (\lambda_l I - A)\, B_{lm_l} &= 0, \\ (\lambda_l I - A)\, B_{lm_{l-1}} &= -B_{lm_l}, \\ &\ \,\vdots \\ (\lambda_l I - A)\, B_{l1} &= -B_{l2}. \end{aligned} \tag{13}$$

If λ_l is a simple eigenvalue, $m_l = 1$ and (13) reduces to (12). In the first of equations (13) the matrix B_{lm_l} either vanishes identically or contains at least one column, which does not vanish. In the latter case we construct a vector X_{lm_l} with that column, so that

$$AX_{lm_l} = \lambda_l X_{lm_l}.$$

Then, from the second equation in (13), we find a column in B_{lm_l-1}, say, X_{lm_l-1}, such that

$$AX_{lm_l-1} = \lambda_l X_{lm_l-1} + X_{lm_l},$$

and by going on to all remaining equations, we obtain a sequence X_{lj} such that

$$\begin{aligned} AX_{lj} &= \lambda_l X_{lj} + X_{lj+1}, \qquad j = 1, 2, \cdots, m_l - 1, \\ AX_{lm_l} &= \lambda_l X_{lm_l}. \end{aligned} \tag{14}$$

Thus we obtain a system of n vectors: $X_{11}, X_{12}, \cdots X_{1m_1}, X_{21} \cdots X_{2m_2} \cdots, X_{k1} \cdots X_{km_k}$. These can be shown to be linearly independent, as in Section 1.6, and now we construct a nonsingular matrix T with these n-vectors as columns. Then we investigate the matrix

$$T^{-1}AT.$$

First we observe that

$$T^{-1}AT \begin{pmatrix} 1 \\ 0 \\ 0 \\ \vdots \\ 0 \end{pmatrix} = T^{-1}AX_{11} = T^{-1}(\lambda_1 X_{11} + X_{12}) = \lambda_1 \begin{pmatrix} 1 \\ 0 \\ 0 \\ \vdots \\ 0 \end{pmatrix} + \begin{pmatrix} 0 \\ 1 \\ 0 \\ \vdots \\ 0 \end{pmatrix}$$

by the use of (14). Furthermore

$$T^{-1}AT \begin{pmatrix} 0 \\ 1 \\ 0 \\ \vdots \\ 0 \end{pmatrix} = T^{-1}AX_{12} = T^{-1}(\lambda_1 X_{12} + X_{13}) = \lambda_1 \begin{pmatrix} 0 \\ 1 \\ 0 \\ \vdots \\ 0 \end{pmatrix} + \begin{pmatrix} 0 \\ 0 \\ 1 \\ \vdots \\ 0 \end{pmatrix},$$

and so on, up to

$$T^{-1}AT \begin{pmatrix} \delta_{1m_1} \\ \delta_{2m_1} \\ \vdots \\ \delta_{nm_1} \end{pmatrix} = T^{-1}AX_{1m_1} = T^{-1}\lambda_1 X_{1m_1} = \lambda_1 \begin{pmatrix} \delta_{1m_1} \\ \delta_{2m_1} \\ \vdots \\ \delta_{nm_1} \end{pmatrix}.$$

One can see, therefore, that

$$T^{-1}AT = \begin{pmatrix} M_1 & & & 0 \\ & M_2 & & \\ & & \ddots & \\ 0 & & & M_k \end{pmatrix}, \tag{15}$$

that is, a matrix whose elements are all zeros except in the blocks along the main diagonal, and

$$M_1 = \begin{pmatrix} \lambda_1 & 0 & 0 & \cdots & 0 \\ 1 & \lambda_1 & 0 & \cdots & 0 \\ 0 & 1 & \lambda_1 & \cdots & \\ \vdots & \vdots & & & \\ 0 & 0 & \cdots\cdots & & \lambda_1 \end{pmatrix},$$

and similarly for the other blocks. That is, the block M_l is of dimension $m_l \times m_l$ and has λ_l in each component on the main diagonal. The diagonal below the main diagonal contains only 1's and all other elements are 0.

EXAMPLE 2. Solve

$$A = \begin{pmatrix} 0 & 1 & 0 \\ 0 & 0 & 1 \\ 2 & -5 & 4 \end{pmatrix}.$$

A calculation shows that the inverse of $(sI - A)$ is

$$B(s) = \frac{1}{(s-1)^2(s-2)} \begin{pmatrix} s^2 - 4s + 5 & s - 4 & 1 \\ 2 & s(s-4) & s \\ 2s & -5s + 2 & s^2 \end{pmatrix}$$

$$= \frac{1}{(s-1)^2} \begin{pmatrix} -2 & 3 & -1 \\ -2 & 3 & -1 \\ -2 & 3 & -1 \end{pmatrix} + \frac{1}{s-1} \begin{pmatrix} 0 & 2 & -1 \\ -2 & 5 & -2 \\ -4 & 8 & -3 \end{pmatrix} + \frac{1}{s-2} \begin{pmatrix} 1 & -2 & 1 \\ 2 & -4 & 2 \\ 4 & -8 & 4 \end{pmatrix}.$$

Here A has the double eigenvalue 1, and the simple eigenvalue 2.

We now let

$$X_{11} = \begin{pmatrix} -1 \\ -2 \\ -3 \end{pmatrix}, \qquad X_{12} = \begin{pmatrix} -1 \\ -1 \\ -1 \end{pmatrix}, \qquad X_{21} = \begin{pmatrix} 1 \\ 2 \\ 4 \end{pmatrix},$$

and of course

$$AX_{11} = X_{11} + X_{12},$$
$$AX_{12} = X_{12},$$
$$AX_{21} = 2X_{21}.$$

Now we construct

$$T = \begin{pmatrix} -1 & -1 & 1 \\ -2 & -1 & 2 \\ -3 & -1 & 4 \end{pmatrix}, \qquad T^{-1} = \begin{pmatrix} 2 & -3 & 1 \\ -2 & 1 & 0 \\ 1 & -2 & 1 \end{pmatrix},$$

and find that

$$T^{-1}AT = \begin{pmatrix} 1 & 0 & 0 \\ 1 & 1 & 0 \\ 0 & 0 & 2 \end{pmatrix},$$

which is (15) for this case.

In the previous discussion it was assumed that in the first of the equations (13) the matrix B_{lm_l} does not vanish identically. If it does vanish, then B_{lm_l-1} again either does or does not vanish identically. Eventually a B_{lr} is found that does not vanish identically, and then one has

$$\begin{aligned} (\lambda_l I - A)\, B_{lr} &= 0, \\ (\lambda_l I - A)\, B_{lr-1} &= -B_{lr}, \\ (\lambda_l I - A)\, B_{l1} &= -B_{l2}. \end{aligned} \tag{16}$$

As before, one can now find column vectors such that

$$\begin{aligned} AX_{lj} &= \lambda_l X_{lj} + X_{lj+1}, \qquad j = 1, 2, \cdots r-1, \\ AX_{lr} &= \lambda_l X_{lr}, \end{aligned} \tag{17}$$

and since λ_l is an eigenvalue of multiplicity m_l, it develops that one can still find $m_l - r$ vectors that satisfy relations similar to (17). In Section 2.8 it will be shown that a $n \times n$ system of type (4) must have precisely n linearly independent solutions. Therefore the n terms of (10) must be linearly independent. It follows that n linearly independent vectors X_{lj} must exist, satisfying relations of the type (17). In this case one find that $T^{-1}AT$ is of the form in (15) except that the M_l block is slightly different from the type described before. From (17) it follows that the main diagonal still contains m_l λ_l terms, but the diagonal below the main diagonal now contains both 1's and 0's.

EXAMPLE 3.

$$A = \begin{pmatrix} 5 & -1 & 1 & 1 & 0 & 0 \\ 1 & 3 & -1 & -1 & 0 & 0 \\ 0 & 0 & 4 & 0 & 1 & 1 \\ 0 & 0 & 0 & 4 & -1 & -1 \\ 0 & 0 & 0 & 0 & 3 & 1 \\ 0 & 0 & 0 & 0 & 1 & 3 \end{pmatrix};$$

$$\Delta(s) = |\, sI - A \,| = (s-4)^5 (s-2).$$

The inverse of $(sI - A)$ is found to be

$$B(s) = \frac{1}{\Delta(s)} \begin{pmatrix} (s-4)^3(s-3)(s-2) & -(s-4)^3(s-2) & (s-4)^2(s-2)^2 & (s-4)^2(s-2)^2 & 0 & 0 \\ (s-4)^3(s-2) & (s-5)(s-4)^3(s-2) & -(s-6)(s-4)^2(s-2) & -(s-6)(s-4)^2(s-2) & 0 & 0 \\ 0 & 0 & (s-4)^4(s-2) & 0 & (s-4)^3(s-2) & (s-4)^3(s-2) \\ 0 & 0 & 0 & (s-4)^4(s-2) & -(s-4)^3(s-2) & -(s-4)^3(s-2) \\ 0 & 0 & 0 & 0 & (s-4)^4(s-3) & (s-4)^4 \\ 0 & 0 & 0 & 0 & (s-4)^4 & (s-4)^4(s-3) \end{pmatrix}$$

$$= \frac{B_{13}}{(s-4)^3} + \frac{B_{12}}{(s-4)^2} + \frac{B_{11}}{(s-4)} + \frac{B_{21}}{s-2},$$

where

$$B_{21} = \begin{pmatrix} 0 & 0 & 0 & 0 & 0 & 0 \\ 0 & 0 & 0 & 0 & 0 & 0 \\ 0 & 0 & 0 & 0 & 0 & 0 \\ 0 & 0 & 0 & 0 & 0 & 0 \\ 0 & 0 & 0 & 0 & \frac{1}{2} & -\frac{1}{2} \\ 0 & 0 & 0 & 0 & -\frac{1}{2} & \frac{1}{2} \end{pmatrix}$$

$$B_{13} = \begin{pmatrix} 0 & 0 & 2 & 2 & 0 & 0 \\ 0 & 0 & 2 & 2 & 0 & 0 \\ 0 & 0 & 0 & 0 & 0 & 0 \\ 0 & 0 & 0 & 0 & 0 & 0 \\ 0 & 0 & 0 & 0 & 0 & 0 \\ 0 & 0 & 0 & 0 & 0 & 0 \end{pmatrix}$$

$$B_{12} = \begin{pmatrix} 1 & -1 & +1 & 1 & 0 & 0 \\ 1 & -1 & -1 & -1 & 0 & 0 \\ 0 & 0 & 0 & 0 & 1 & 1 \\ 0 & 0 & 0 & 0 & -1 & -1 \\ 0 & 0 & 0 & 0 & 0 & 0 \\ 0 & 0 & 0 & 0 & 0 & 0 \end{pmatrix}$$

$$B_{11} = \begin{pmatrix} 1 & 0 & 0 & 0 & 0 & 0 \\ 0 & 1 & 0 & 0 & 0 & 0 \\ 0 & 0 & 1 & 0 & 0 & 0 \\ 0 & 0 & 0 & 1 & 0 & 0 \\ 0 & 0 & 0 & 0 & \frac{1}{2} & \frac{1}{2} \\ 0 & 0 & 0 & 0 & \frac{1}{2} & \frac{1}{2} \end{pmatrix}$$

We now let

$$X_3 = \begin{pmatrix} 2 \\ 2 \\ 0 \\ 0 \\ 0 \\ 0 \end{pmatrix} \qquad X_2 = \begin{pmatrix} 1 \\ -1 \\ 0 \\ 0 \\ 0 \\ 0 \end{pmatrix} \qquad X_1 = \begin{pmatrix} 0 \\ 0 \\ 1 \\ 0 \\ 0 \\ 0 \end{pmatrix}$$

$$X_5 = \begin{pmatrix} 0 \\ 0 \\ 1 \\ -1 \\ 0 \\ 0 \end{pmatrix}, \qquad X_4 = \begin{pmatrix} 0 \\ 0 \\ 0 \\ 0 \\ \frac{1}{2} \\ \frac{1}{2} \end{pmatrix}, \qquad X_6 = \begin{pmatrix} 0 \\ 0 \\ 0 \\ 0 \\ -\frac{1}{2} \\ +\frac{1}{2} \end{pmatrix}$$

and see that

$$\begin{aligned}
AX_3 &= 4X_3,\\
AX_2 &= 4X_2 + X_3,\\
AX_1 &= 4X_1 + X_2,\\
AX_5 &= 4X_5,\\
AX_4 &= 4X_4 + X_5,\\
AX_6 &= 2X_6.
\end{aligned}$$

By forming the matrix T with columns X_i, one sees from the above relations, or by a direct calculation, that

$$T^{-1}AT = \begin{pmatrix} 4 & 0 & 0 & 0 & 0 & 0\\ 1 & 4 & 0 & 0 & 0 & 0\\ 0 & 1 & 4 & 0 & 0 & 0\\ 0 & 0 & 0 & 4 & 0 & 0\\ 0 & 0 & 0 & 1 & 4 & 0\\ 0 & 0 & 0 & 0 & 0 & 2 \end{pmatrix}.$$

The matrix (15) is known as the Jordan canonical form. It is useful in solving differential equations. The following example will illustrate the point.

$$X' = AX,$$

where A is the matrix of Example 3.

Solution. Let

$$X = TY.$$

Then

$$Y' = T^{-1}ATY,$$

and rewriting this in scalar form, we have

$$\begin{aligned}
y_1' &= 4y_1,\\
y_2' &= 4y_2 + y_1,\\
y_3' &= 4y_3 + y_2,\\
y_4' &= 4y_4,\\
y_5' &= 4y_5 + y_4,\\
y_6' &= 2y_6.
\end{aligned}$$

One can now calculate readily:

$$y_1 = c_1 e^{4z},$$

$$y_2 = c_2 e^{4z} + c_1 z e^{4z},$$

$$y_3 = c_3 e^{4z} + c_2 z e^{4z} + \frac{c_1}{2} z^2 e^{4z},$$

$$y_4 = c_4 e^{4z},$$

$$y_5 = c_5 e^{4z} + c_4 z e^{4z},$$

$$y_6 = c_6 e^{2z},$$

from which X can then be calculated.

2.7 INHOMOGENEOUS SYSTEMS OF LINEAR DIFFERENTIAL EQUATIONS

In the previous sections it was shown that one can always obtain a general solution of a linear homogeneous system with constant coefficients. We now turn our attention to inhomogeneous systems of the form

$$X' = A(z)\,X + F(z),$$

with (1)

$$X(0) = X_0.$$

With this system we associate a second system of equations, known as the adjoint system, namely

$$Y' = -A^T(z)\,Y, \tag{2}$$

where A^T denotes the transpose of A. This system and its application in the subsequent paragraphs should be compared to the adjoint equation and its application in Section 2.2.

We now suppose that a solution of the homogeneous system (2) has been obtained, say, Y_1. Then we find that

$$(Y_1^T X)' = Y_1^T X' + Y_1^{T\prime} X = Y_1^T(AX + F) - Y_1^T AX = Y_1^T F$$

is a first-order equation, and since the right side is known, one can integrate:

$$Y_1^T X = Y_1^T(0)\,X_0 + \int_0^z Y_1^T(t) F(t)\,dt,$$

or in scalar form, if

$$Y_1^T = (y_{11} y_{12} \cdots y_{1n}),$$

$$\sum_{k=1}^n y_{1k} x_k = Y_1^T(0)\,X_0 + \int_0^z Y_1^T(t) F(t)\,dt. \tag{3}$$

This is a linear relationship among the x_i. One can now eliminate one of the x_i from (1), thereby reducing the order of the system from n to $n - 1$. Similarly, if a second linearly independent solution of (2) can be found (say, Y_2) where

$$Y_2^T = (y_{21} y_{22} \cdots y_{2n}),$$

then one has

$$\sum_{k=1}^{n} y_{2k} x_k = Y_2^T(0)\, X_0 + \int_0^z Y_2^T(t) F(t)\, dt. \tag{4}$$

By the use of (3) and (4) one could reduce the system from one of order n to one of order $n - 2$. More generally, if l linearly independent solutions of (2) can be found, the order of (1) can be reduced to $n - l$. If a complete system of n linearly independent solutions of (2) is found, one obtains

$$\sum_{k=1}^{n} y_{ik} x_k = Y_i^T(0)\, X_0 + \int_0^z Y_i^T(t) F(t)\, dt, \qquad i = 1, 2, \cdots n. \tag{5}$$

The system (5) is now an algebraic system of n equations in n unknowns. All that is left now to solve (1) is to solve (5). If the matrix

$$\Phi(z) = \begin{pmatrix} y_{11} & y_{12} & \cdots & y_{1n} \\ y_{21} & y_{22} & \cdots & y_{2n} \\ \vdots & & & \\ y_{n1} & y_{n2} & & y_{nn} \end{pmatrix}$$

is defined, the system (5) can be rewritten as

$$\Phi(z)\, X = \Phi(0)\, X_0 + \int_0^z \Phi(t) F(t)\, dt. \tag{6}$$

That Φ has an inverse follows from the fact that, by hypothesis, its rows are transposes of linearly independent solutions of (2). Then the solution of (1) becomes

$$X = \Phi^{-1}(z)\, \Phi(0)\, X_0 + \Phi^{-1}(z) \int_0^z \Phi(t) F(t)\, dt. \tag{7}$$

Evidently, as follows from (2), Φ^T satisfies the differential equation

$$\Phi^{T'} = -A^T \Phi^T,$$

or by taking transposes,

$$\Phi' = -\Phi A. \tag{8}$$

It is easy to rederive (6) and (7) directly once a nonsingular matrix Φ satisfying (8) has been found, for then

$$(\Phi X)' = \Phi X' + \Phi' X = \Phi(AX + F) - (\Phi A)\, X = \Phi F,$$

from which we find by integration

$$\Phi X = \Phi(0)\, X_0 + \int_0^z \Phi(t) F(t)\, dt,$$

which is, of course, (6).

In the case of a system with a constant matrix A a solution of (8) can be written down by inspection, namely,

$$\Phi = \Phi_0 e^{-Az}$$

where Φ_0 is an arbitrary nonsingular matrix. Evidently

$$\Phi^{-1} = e^{Az}\Phi_0^{-1},$$

so that (7) becomes

$$\begin{aligned} X &= e^{Az}\Phi_0^{-1}\Phi_0 X_0 + e^{Az}\Phi_0^{-1} \int_0^z \Phi_0 e^{-At} F(t)\, dt \\ &= e^{Az} X_0 + \int_0^z e^{A(z-t)} F(t)\, dt. \end{aligned} \tag{9}$$

One can, of course, check by differentiation that (9) is a solution of (1), provided A is a constant matrix.

Example 1. Solve

$$X' = \begin{pmatrix} 0 & 1 & 0 \\ 0 & 0 & 1 \\ -6 & -11 & -6 \end{pmatrix} X + \begin{pmatrix} 0 \\ 0 \\ e^{-t} \end{pmatrix},$$

$$X(0) = \begin{pmatrix} 0 \\ 0 \\ 0 \end{pmatrix}$$

Solution. The adjoint system corresponding to this system is

$$Y' = \begin{pmatrix} 0 & 0 & 6 \\ -1 & 0 & 11 \\ 0 & -1 & 6 \end{pmatrix} Y.$$

We now find by the methods investigated in Section 2.5 that

$$Y_1 = e^z \begin{pmatrix} 6 \\ 5 \\ 1 \end{pmatrix}, \qquad Y_2 = e^{2z} \begin{pmatrix} 3 \\ 4 \\ 1 \end{pmatrix}, \qquad Y_3 = e^{3z} \begin{pmatrix} 2 \\ 3 \\ 1 \end{pmatrix},$$

so that

$$\Phi(z) = \begin{pmatrix} 6e^z & 5e^z & e^z \\ 3e^{2z} & 4e^{2z} & e^{2z} \\ 2e^{3z} & 3e^{3z} & e^{3z} \end{pmatrix},$$

and

$$\Phi(0) = \begin{pmatrix} 6 & 5 & 1 \\ 3 & 4 & 1 \\ 2 & 3 & 1 \end{pmatrix},$$

and

$$\Phi^{-1}(z) = \begin{pmatrix} \frac{1}{2}e^{-z} & -e^{-2z} & \frac{1}{2}e^{-3z} \\ -\frac{1}{2}e^{-z} & 2e^{-2z} & -\frac{3}{2}e^{-3z} \\ \frac{1}{2}e^{-z} & -4e^{-2z} & \frac{9}{2}e^{-3z} \end{pmatrix}.$$

It now follows that

$$\Phi' = -\Phi A,$$

and

$$\Phi^{T'} = -A^T\Phi^T,$$

so that

$$(\Phi X)' = \Phi X' + \Phi' X = \Phi(AX + F) - \Phi AX = \Phi F$$

and

$$\Phi(z)\,X = \int_0^z \Phi(t)F(t)\,dt = \int_0^z \begin{pmatrix} 6e^t & 5e^t & e^t \\ 3e^{2t} & 4e^{2t} & e^{2t} \\ 2e^{3t} & 3e^{3t} & e^{3t} \end{pmatrix} \begin{pmatrix} 0 \\ 0 \\ e^{-t} \end{pmatrix} dt$$

$$= \int_0^z \begin{pmatrix} 1 \\ e^t \\ e^{2t} \end{pmatrix} dt = \begin{pmatrix} z \\ e^z - 1 \\ \dfrac{e^{2z}}{2} - \dfrac{1}{2} \end{pmatrix}$$

$$X = \begin{pmatrix} \frac{1}{2}e^{-z} & -e^{-2z} & \frac{1}{2}e^{-3z} \\ -\frac{1}{2}e^{-z} & 2e^{-2z} & -\frac{3}{2}e^{-3z} \\ \frac{1}{2}e^{-z} & -4e^{-2z} & \frac{9}{2}e^{-3z} \end{pmatrix} \begin{pmatrix} z \\ e^z - 1 \\ \frac{1}{2}(e^{2z} - 1) \end{pmatrix}$$

$$= \begin{pmatrix} \frac{1}{2}ze^{-z} & -\frac{3}{4}e^{-z} & +e^{-2z} & -\frac{1}{4}e^{-3z} \\ -\frac{1}{2}ze^{-z} & +\frac{5}{4}e^{-z} & -2e^{-2z} & +\frac{3}{4}e^{-3z} \\ \frac{1}{2}ze^{-z} & -\frac{7}{4}e^{-z} & +4e^{-2z} & -\frac{9}{4}e^{-3z} \end{pmatrix},$$

which is, as can be verified by a direct calculation, the required solution.

EXAMPLE 2. Solve

$$X' = \begin{pmatrix} 0 & \dfrac{4z^2 - 2z + 2}{2z - 1} \\ -1 & -\dfrac{4z^2 + 1}{2z - 1} \end{pmatrix} X + \begin{pmatrix} e^{-z} \\ 0 \end{pmatrix},$$

$$X(0) = \begin{pmatrix} 0 \\ 0 \end{pmatrix}.$$

Solution. The adjoint system is given by

$$Y' = \begin{pmatrix} 0 & 1 \\ -\dfrac{4z^2 - 2z + 2}{2z - 1} & \dfrac{4z^2 + 1}{2z - 1} \end{pmatrix} Y.$$

One can verify that a particular solution is given by

$$Y_1 = e^z \begin{pmatrix} 1 \\ 1 \end{pmatrix}.$$

Then

$$(Y_1^T X)' = Y_1^T X' + Y_1^{T\prime} X = Y_1^T(AX + F) - Y_1^T AX = Y_1^T F = Y_1^T \begin{pmatrix} e^{-z} \\ 0 \end{pmatrix} = 1,$$

so that

$$Y_1^T X = z$$

which, when written in scalar form, leads to

$$x_1 + x_2 = ze^{-z}.$$

Elimination of x_1 from the system leads to an equation for x_2, namely,

$$x_2' = -\frac{4z^2 - 2z + 2}{2z - 1} x_2 - ze^{-z}.$$

This equation can be solved by the methods of Section 2.2, and one now finds that

$$x_2 = \frac{-e^{-z^2}}{2z - 1} \int_0^z e^{t^2 - t}(2t^2 - t)\, dt,$$

and

$$x_1 = ze^{-z} + \frac{e^{-z^2}}{2z - 1} \int_0^z e^{t^2 - t}(2t^2 - t)\, dt.$$

The chief advantage of introducing the adjoint system (2) is that once one solution of (2) is known, the order of system (1) can be lowered. But when it is possible to obtain n linearly independent solutions of (2), there is an alternative approach available, which is sometimes more convenient in practice. In the final answer (7), both Φ and Φ^{-1} appear. Rather than first solving (8) to obtain Φ, we shall try to find a differential equation for Φ^{-1}. To do so, let

$$\Phi^{-1} = \Psi.$$

Then

$$\Phi\Psi = I,$$

and by differentiation of this equation and use of (8) it is found that

$$\Psi' = -\Phi^{-1}\Phi'\Psi = -\Psi\Phi'\Psi = -\Psi(-\Phi A\Psi) = (\Psi\Phi) A\Psi = A\Psi. \quad (10)$$

Thus it is seen that the matrix Ψ satisfies a differential equation similar to (1) but homogeneous. We suppose that we now have found a nonsingular matrix that satisfies (10), and we seek to solve (1) under the assumption that

$$X = \Psi Y. \quad (11)$$

Insertion of this in (1) leads to

$$\Psi Y' = F,$$

and solving for Y',

$$Y' = \Psi^{-1}F. \quad (12)$$

From (11) it follows that

$$X = \Psi(z)\, Y_0 + \Psi(z) \int_0^z \Psi^{-1}(t) F(t)\, dt, \quad (13)$$

which is, of course, completely analogous to (7). If initial conditions on X are given, Y_0 can be obtained from (11).

The above method is known as the method of variation of parameters.

2.8 GENERAL THEORY OF LINEAR DIFFERENTIAL EQUATIONS

In the previous sections it has been shown that the general first-order inhomogeneous system,

$$\begin{aligned} X' &= A(z)\, X + F(z), \\ X(0) &= X_0, \end{aligned} \quad (1)$$

can be solved, provided a general solution of the associated adjoint system,

$$Y' = -A^T(z)\, Y, \quad (2)$$

can be found. Furthermore it was shown that a solution of a homogeneous system of type (2) can always be found whenever the matrix A is constant. We shall now consider more general cases, and prove that under certain mild restrictions on A, homogeneous systems with initial conditions have unique solutions. First we shall show that if the equation

$$X' = A(z)\,X \tag{3}$$

has n linearly independent solutions near a point z_0, then they will remain independent in every interval $a < z < b$ including the point z_0, in which $A(z)$ has no singularities; that is, none of the $a_{ij}(z)$ has singularities.

Suppose $X_1, X_2, \cdots X_n$ represent a system of n linearly independent solutions. Let

$$\Phi = (X_1 X_2 \cdots X_n) \tag{4}$$

be a matrix with these n vectors as columns, and let

$$W(z) = \det \Phi.$$

This determinant is traditionally known as the Wronskian. By hypothesis, $W(z_0) \neq 0$. In order to obtain an explicit formula for W, it is most convenient to show first that it satisfies a first-order linear differential equation and then to solve this equation. We write

$$W = \begin{vmatrix} x_{11} & x_{12} & \cdots & x_{1n} \\ x_{21} & x_{22} & \cdots & x_{2n} \\ \vdots & \vdots & & \vdots \\ x_{n1} & x_{n2} & & x_{nn} \end{vmatrix}$$

and differentiate this determinant. A determinant is a sum of $n!$ terms, each of which is a product of n terms. The derivative of a product of n terms is a sum of n terms in each of which the derivative of only one of the terms occurs. From these considerations it follows that

$$W' = \begin{vmatrix} x'_{11} & x'_{12} & \cdots & x'_{1n} \\ x_{21} & x_{22} & & x_{2n} \\ \vdots & \vdots & & \vdots \\ x_{n1} & x_{n2} & & x_{nn} \end{vmatrix} + \begin{vmatrix} x_{11} & x_{12} & \cdots & x_{1n} \\ x'_{21} & x'_{22} & & x'_{2n} \\ \vdots & \vdots & & \vdots \\ x_{n1} & x_{n2} & & x_{nn} \end{vmatrix}$$

$$+ \cdots + \begin{vmatrix} x_{11} & x_{12} & \cdots & x_{1n} \\ x_{21} & x_{22} & \cdots & x_{2n} \\ \vdots & \vdots & & \vdots \\ x'_{n1} & x'_{n2} & \cdots & x'_{nn} \end{vmatrix}. \tag{5}$$

The first term in (5) becomes, after replacing terms of the first row by their corresponding terms from (3),

$$\begin{vmatrix} \sum_{i=1}^{n} a_{1i}(z)\, x_{i1} & \sum_{i=1}^{n} a_{1i}(z)\, x_{i2} & \cdots & \sum_{i=1}^{n} a_{1i}(z)\, x_{in} \\ x_{21} & x_{22} & \cdots & x_{2n} \\ \vdots & \vdots & & \vdots \\ x_{n1} & x_{n2} & \cdots & x_{nn} \end{vmatrix}.$$

If one now subtracts a_{12} times the second row from the first row, a_{13} times the third row from the first, and so on, one obtains

$$\begin{vmatrix} a_{11}x_{11} & a_{11}x_{12} & \cdots & a_{11}x_{11} \\ x_{21} & x_{22} & \cdots & x_{2n} \\ \vdots & \vdots & & \vdots \\ x_{n1} & x_{n2} & & x_{nn} \end{vmatrix} = a_{11} W.$$

After performing similar operations on the remaining terms in (5), one finds that

$$W' = (a_{11} + a_{22} + \cdots + a_{nn})\, W \tag{6}$$

with the initial value

$$W(z_0) = W_0$$

The coefficient of W in (6) is just the sum of the diagonal elements of A and is called the trace of A. The symbol associated with it is tr A. Then (6) becomes

$$W' = \operatorname{tr} AW,$$
$$W(z_0) = W_0,$$

whose solution is

$$W = W_0 \exp\left(\int_{z_0}^{z} \operatorname{tr} A\, dt\right)$$

Such an exponential function can vanish only if the exponent approaches $-\infty$. But this can happen only at a point where one of the terms a_{ii} has a singularity.

Next we show that (3) has a solution with initial vector $X(0) = X_0$, provided the matrix $A(z)$ is a continuous function in z. Rather than consider the vector differential equation (3), we shall consider the matrix differential equation

$$\begin{aligned} \Phi' &= A\Phi, \\ \Phi(0) &= I, \end{aligned} \tag{7}$$

where Φ is defined in (4). Evidently, once such a Φ is found, the solution of (3) is

$$X = \Phi X_0.$$

To show that (7) has a unique solution, we proceed as follows: We define an infinite sequence of matrices $\Phi_0, \Phi_1, \Phi_2, \cdots$ such that

$$\begin{aligned} \Phi_0 &= I, \\ \Phi_1 &= I + \int_0^z A(t)\,\Phi_0(t)\,dt, \\ \Phi_2 &= I + \int_0^z A(t)\,\Phi_1(t)\,dt, \\ &\vdots \\ \Phi_{k+1} &= I + \int_0^z A(t)\,\Phi_k(t)\,dt. \end{aligned} \tag{8}$$

If this sequence has a uniform limit, then its limit function can be expected to satisfy the equation

$$\Phi = I + \int_0^z A(t)\,\Phi(t)\,dt,$$

and by differentiation we see that

$$\Phi' = A\Phi,$$

and also at $z = 0$,

$$\Phi(0) = I.$$

From a simple subtraction we see that

$$\Phi_{k+1} - \Phi_k = \int_0^z A(t)[\Phi_k(t) - \Phi_{k-1}(t)]\,dt$$

and

$$\begin{aligned} \| \Phi_{k+1} - \Phi_k \| &= \left\| \int_0^z A(t)[\Phi_k(t) - \Phi_{k-1}]\,dt \right\| \\ &\leqslant \int_0^z \| A(t) \|\,\| \Phi_k(t) - \Phi_{k-1}(t) \|\,dt. \end{aligned} \tag{9}$$

If we restrict ourselves to a closed and bounded interval $0 \leqslant z \leqslant z_1$ in which $A(z)$ is continuous, then $\| A(z) \|$ is bounded so that there exists a positive integer M such that

$$\| A(z) \| \leqslant M \qquad \text{for } 0 \leqslant z \leqslant z_1.$$

In that case, after setting

$$\| \Phi_k - \Phi_{k-1} \| = u_k,$$

one obtains from (9),

$$u_{k+1} \leqslant M \int_0^z u_k \, dt. \tag{10}$$

Evidently

$$u_1 = \| \Phi_1 - \Phi_0 \| = \left\| \int_0^z A(t) \, \Phi_0(t) \, dt \right\| \leqslant \int_0^z \| A(t) \| \, \| \Phi_0(t) \| \, dt \leqslant Mnz,$$

since

$$\| \Phi_0(t) \| = \| I \| = n.$$

Then, from (10),

$$u_2 \leqslant M \int_0^z u_1 \, dt = M^2 n \int_0^z t \, dt = \frac{nM^2 z^2}{2},$$

$$u_3 \leqslant M \int_0^z u_2 \, dt = \frac{nM^3}{2} \int_0^z t^2 \, dt = \frac{nM^3 z^3}{3!},$$

and proceeding inductively, one finds that

$$u_k \leqslant \frac{nM^k z^k}{k!}. \tag{11}$$

Next we use the obvious identity

$$\Phi_k = \Phi_0 + (\Phi_1 - \Phi_0) + (\Phi_2 - \Phi_1) + \cdots + (\Phi_k - \Phi_{k-1}) \tag{12}$$

and test the series for convergence. Evidently

$$\begin{aligned} \| \Phi_k \| &\leqslant \| \Phi_0 \| + \| \Phi_1 - \Phi_0 \| + \cdots + \| \Phi_k - \Phi_{k-1} \| \\ &\leqslant n + u_1 + u_2 + \cdots + u_k \\ &\leqslant n \left(1 + Mz + \frac{M^2 z^2}{2!} + \cdots + \frac{M^k z^k}{k!} \right). \end{aligned}$$

The series on the right is a convergent series and

$$\lim_{k \to \infty} \| \Phi_k \| \leqslant ne^{Mz}.$$

Therefore the series (12) converges so that Φ_k has a limit. This establishes the fact that the iteration process (8) converges and (7) has a solution. To show that this solution is unique, we now assume that (7) has a second solution Ψ.

In analogy to (9) we find that

$$\| \Phi - \Psi \| \leqslant \int_0^z \| A(t) \| \, \| \Phi - \Psi \| \, dt \leqslant M \int_0^z \| \Phi - \Psi \| \, dt.$$

From the preceding inequality one can show that $\| \Phi - \Psi \|$ must vanish identically. First we integrate both sides to obtain

$$\int_0^z \| \Phi - \Psi \| \, dt \leqslant M \int_0^z dz \int_0^z \| \Phi - \Psi \| \, dt = M \int_0^z (z - t) \| \Phi - \Psi \| \, dt,$$

so that

$$\| \Phi - \Psi \| \leqslant M^2 \int_0^z (z - t) \| \Phi - \Psi \| \, dt.$$

By a second integration we find that

$$\| \Phi - \Psi \| \leqslant M^3 \int_0^z \frac{(z - t)^2}{2!} \| \Phi - \Psi \| \, dt,$$

and after repeated integrations,

$$\| \Phi - \Psi \| \leqslant \frac{M^{k+1}}{k!} \int_0^z (z - t)^k \| \Phi - \Psi \| \, dt.$$

Due to the $k!$ in the denominator, it follows that as k approaches infinity,

$$\| \Phi - \Psi \| \leqslant 0,$$

from which we deduce that

$$\Phi = \Psi.$$

We can now summarize the previous results in the following theorem.

THEOREM: The equation

$$X' = A(z) X$$
$$X(0) = X_0,$$

will have a unique solution in every z interval containing the origin, providing the matrix $A(z)$ is continuous in that interval.

Furthermore, if $X_1, X_2, \cdots, X_n$ represent a system of n linearly independent solutions, then the determinant

$$W = \det (X_1 X_2 \cdots X_n)$$

is given explicitly by

$$W = W_0 \exp \int_0^z \operatorname{tr} A \, dt.$$

If $W_0 \neq 0$, W will not vanish for any value of z where tr A has no singularity.

One can deduce one more important theorem by the methods of this section.

THEOREM: The solutions of the differential equation depend continuously on the initial data. That is, if Φ and Ψ are two matrix solutions of the equation

$$\Phi' = A(z)\,\Phi$$

such that

$$\| \Phi(0) - \Psi(0) \| < \epsilon$$

then

$$\| \Phi - \Psi \| \leqslant \delta(\epsilon) \qquad \text{for } 0 \leqslant z \leqslant z_1$$

and

$$\lim_{\epsilon \to 0} \delta(\epsilon) = 0.$$

Proof. Evidently

$$\Phi = \Phi(0) + \int_0^z A(t)\,\Phi(t)\,dt,$$

$$\Psi = \Psi(0) + \int_0^z A(t)\,\Psi(t)\,dt,$$

$$\Phi - \Psi = \Phi(0) - \Psi(0) + \int_0^z A(t)[\Phi(t) - \Psi(t)]\,dt,$$

so that

$$\| \Phi - \Psi \| \leqslant \| \Phi(0) - \Psi(0) \| + \int_0^z \| A(t) \|\,\| \Phi(t) - \Psi(t) \|\,dt$$

$$\leqslant \epsilon + M \int_0^z \| \Phi(t) - \Psi(t) \|\,dt,$$

where

$$\| A(t) \| \leqslant M \qquad \text{for } 0 \leqslant t < z_1.$$

Integrating both sides of the inequality, one obtains

$$\int_0^z \| \Phi(t) - \Psi(t) \|\,dt \leqslant \epsilon z + M \int_0^z (z - t)\,\| \Phi(t) - \Psi(t) \|\,dt,$$

so that, by combining the last two inequalities,

$$\| \Phi - \Psi \| \leqslant \epsilon + M\epsilon z + M^2 \int_0^z (z - t)\,\| \Phi(t) - \Psi(t) \|\,dt.$$

By repeating this process k times, one finds

$$\| \Phi - \Psi \| \leqslant \epsilon \left[1 + Mz + \frac{M^2 z^2}{2!} + \cdots + \frac{M^k z^k}{k!} \right] + \frac{M^{k+1}}{k!} \int_0^z (z - t)^k \| \Phi(t) - \Psi(t) \| \, dt,$$

and in the limit as k approaches infinity,

$$\| \Phi - \Psi \| \leqslant \epsilon e^{Mz}.$$

Evidently an acceptable value of δ will be

$$\delta(\epsilon) = 2\epsilon e^{Mz_1},$$

completing the proof of the theorem.

Problems

1. Solve the following first-order differential equations:

(a) $y' - y = 0, \qquad y(0) = 1.$

(b) $xy' + 5y = 6x^2, \qquad y(1) = 3.$

(c) $x^3 y' + 2y = x^3 + 2x, \qquad y(1) = e + 1.$

(d) $y' + y = x, \qquad y(0) = 0.$

2. Solve the following differential equation and discuss the uniqueness of the solution; that is, for which domain including the initial point is the solution unique? What happens if that domain is enlarged?

$$(\tan x)\, y' - 2y = 0,$$

$$y\left(\frac{\pi}{2}\right) = 1.$$

3. Solve the following systems of equations:

(a) $X' = \begin{pmatrix} 2 & 1 \\ 3 & 4 \end{pmatrix} X, \qquad X(0) = \begin{pmatrix} 2 \\ 2 \end{pmatrix}.$

(b) $X' = \begin{pmatrix} 5 & 1 \\ 4 & 2 \end{pmatrix} X, \qquad X(0) = \begin{pmatrix} 5 \\ 0 \end{pmatrix}.$

(c) $X' = \begin{pmatrix} 4 & 2 & -2 \\ -5 & 3 & 2 \\ -2 & 4 & 1 \end{pmatrix} X, \qquad X(0) = \begin{pmatrix} 3 \\ 0 \\ 4 \end{pmatrix}.$

4. Solve the following differential equations by first writing them as systems and then solving the equation:

(a) $y'' - 3y' - 10y = 0,$
$y(0) = 3, \quad y'(0) = 15.$

(b) $y'' + y' - 12y = 0,$
$y(0) = 0, \quad y'(0) = 21.$

(c) $y''' - 7y'' + 2y' + 40y = 0,$
$y(0) = 3, \quad y'(0) = 7, \quad y''(0) = 27.$

(d) $y''' - 3y'' + y' + 3y = 0,$
$y(0) = 0, \quad y'(0) = 8, \quad y''(0) = 24.$

(e) $y''' - y'' + 4y' - 4y = 0,$
$y(0) = 1, \quad y'(0) = 0, \quad y''(0) = 0.$

5. Show that if A is a matrix, then

(a) $e^{c_1A+c_2A} = e^{c_1A}e^{c_2A}.$

(b) $(e^A)^{-1} = e^{-A}.$

(c) $(e^A)^m = e^{mA}, \quad (m \text{ an integer}).$

(d) $e^0 = I.$

6. What are the conditions on A and B so that $e^{A+B} = e^Ae^B$?

7. Derive by integration the Laplace transforms of the following functions:

(a) $\sin az.$

(b) $\cos az.$

(c) $e^{az} \sin bz.$

(d) $e^{az} \cos bz.$

(e) $z^4 + 4.$

(f) $(1 - z^2)\, e^{2z}.$

8. Let $\mathscr{L}(y)$ denote the Laplace transform of y. Show that

$$\mathscr{L}(y') = s\mathscr{L}(y) - y(0),$$

and more generally,

$$\mathscr{L}(y^{(n)}) = s^n\mathscr{L}(y) - y^{(n-1)}(0) - \cdots s^{n-1}y(0).$$

9. Use the method of the Laplace transform to solve the following differential equations:

(a) $y''' - 9y'' + 24y' - 16y = 0,$
$y(0) = 0, \quad y'(0) = -11, \quad y''(0) = -61.$

(b) $X' = \begin{pmatrix} 0 & 0 & 1 \\ -2 & -3 & -9 \\ 0 & 2 & 6 \end{pmatrix} X, \quad X(0) = \begin{pmatrix} -2 \\ -5 \\ 0 \end{pmatrix}.$

(c) $X' = \begin{pmatrix} 3 & 5 & -1 \\ 0 & -2 & 1 \\ 1 & 0 & 0 \end{pmatrix} X, \qquad X(0) = \begin{pmatrix} -17 \\ 1 \\ -5 \end{pmatrix}.$

(d) $X' = \begin{pmatrix} 4 & 1 \\ -1 & 2 \end{pmatrix} X.$

10. Show that the vectors X_{lj}, which satisfy (14) of Section 2.6, are linearly independent.

11. Show that the Jordan canonical form (15) of Section 2.6 is unique, in the sense that the number of blocks M_i and their respective size are unique. Use the uniqueness of the solution of the differential equation.

12. Let A be a constant matrix. A and A^T have the same eigenvalues. Consider

$$X' = AX,$$
$$Y' = -A^T Y,$$

and the solutions

$$X = e^{\lambda_i z} C_i,$$
$$Y = e^{-\lambda_j z} Dj.$$

Show that if $\lambda_i \neq \lambda_j$, then $D_j^T C_i = 0$.

13. Use the method of the preceding problem to show that if A is a real symmetric matrix and

$$X_1 = e^{\lambda_1 z} C_1, \qquad X_2 = e^{\lambda_2 z} C_2,$$

are solutions of

$$X' = AX,$$

then, if

$$\lambda_1 \neq \lambda_2, \qquad C_1^T C_2 = 0.$$

14. Show that if A is a real symmetric matrix, then its Jordan canonical form is a pure diagonal matrix. (*Hint*: Use the idea underlying the preceding problem.)

15. Consider the following three matrices

(a) $\begin{pmatrix} 1 & 0 & 1 \\ 0 & 2 & 0 \\ 0 & 0 & 1 \end{pmatrix}.$

(b) $\begin{pmatrix} 1 & 1 & 2 \\ 0 & 2 & 0 \\ 1 & 4 & 0 \end{pmatrix}.$

(c) $\begin{pmatrix} 0 & -8 & 4 \\ 0 & 2 & 0 \\ 2 & 3 & -2 \end{pmatrix}.$

For each of these determine $B(s) = (sI - A)^{-1}$. Determine T such that $T^{-1}AT$ is in Jordan canonical form. Verify by multiplication that $T^{-1}AT$ has the correct form.

16. Solve the following differential equations:

(a) $$X' = \begin{pmatrix} 0 & 0 & 3 \\ 3 & 2 & 0 \\ 1 & 1 & -1 \end{pmatrix} X, \qquad X(0) = \begin{pmatrix} -3 \\ -9 \\ 13 \end{pmatrix}.$$

(b) $y''' + 3y'' + 3y' + y = 0,$
$y(0) = 0, \quad y'(0) = 1, \quad y''(0) = -2.$

(c) $y''' - 3y'' + 4y = 0,$
$y(0) = 3, \quad y'(0) = 6, \quad y''(0) = 12.$

(d) $y' + y = e^{-x}, \quad y(0) = 1.$

(e) $y'' + 2y' + 2y = \sin z,$
$y(0) = 0, \quad y'(0) = 1.$

(f) $y''' + 2y'' + y' + 2y = -2e^z,$
$y(0) = 1, \quad y'(0) = 1, \quad y''(0) = 1.$

17. The solutions of the differential equation,

$$zy'' + (1 - z)\,y' + ny = 0,$$

where n is an integer, which are polynomials, are known as Laguerre polynomials. They are given in general by

$$L_n(z) = e^z \left(\frac{d}{dz}\right)^n z^n e^{-z}.$$

Verify the above for $n = 1, 2, 3$, and find the complete solution of the equation by reduction of order.

18. Show that the Wronskian of the nth-order differential equation

$$y^{(n)} + a_1(z)\,y^{(n-1)}(z) + \cdots + a_n(z)\,y(z) = 0$$

is given by

$$W = c \exp\left(-\int_0^z a_1(t)\,dt\right)$$

19. Solve the equation

$$zy'' + (z - 2)\,y' - 3y = 0.$$

$y = z^3$ is a solution.

3 Series Solutions for Linear Differential Equations

3.1 DIFFERENTIAL EQUATIONS WITH ANALYTIC COEFFICIENTS

In the preceding chapter equations of the form

$$X' = A(z)X + F(z),$$
$$X(0) = X_0 \tag{1}$$

were investigated. It was shown that whenever A was a constant matrix, a solution could be obtained by an explicit calculation. In any case, once a complete solution of the adjoint equation had been obtained, the inhomogeneous equation (1) could be solved. But no explicit methods were discussed for homogeneous equations with nonconstant coefficients.

Generally speaking, it is impossible to obtain closed form solutions for differential equations. One has to resort to a variety of other tools to obtain information regarding the nature of the solution. One method, which is particularly simple, applies to equations with analytic coefficients; that is, those cases where the matrix A can be expanded in a convergent Taylor series about some point (say, z_0),

$$A(z) = \sum_{k=0}^{\infty} A_k(z - z_0)^k, \tag{2}$$

and the A_k are constant matrix coefficients.

Without loss of generality we can always shift our coordinate axis so as to make $z_0 = 0$. We now turn our attention to the equation

$$X' = \sum_{k=0}^{\infty} A_k z^k X,$$
$$X(0) = X_0 . \tag{3}$$

It seems plausible to assume a solution of the form

$$X = \sum_{k=0}^{\infty} C_k z^k, \tag{4}$$

where the C_k are unknown vector coefficients. To determine these unknowns, one inserts this trial solution in (3) and obtains

$$\sum_{k=0}^{\infty} kC_k z^{k-1} = \sum_{k=0}^{\infty} A_k z^k \sum_{k=0}^{\infty} C_k z^k. \tag{5}$$

Next one carries out the multiplication on the right and rearrranges the series so that

$$\sum_{k=0}^{\infty} kC_k z^{k-1} = \sum_{k=0}^{\infty} z^k \sum_{l=0}^{k} A_{k-l} C_l. \tag{6}$$

It is well known from the theory of power series that two power series are identical if and only if the coefficients of corresponding power terms are equal. This theorem can be easily extended to cover cases in which the coefficients in the series are vector terms, as in (6). One can therefore equate the coefficients of z^k on both sides of (6) to obtain

$$(k+1)\, C_{k+1} = \sum_{l=0}^{k} A_{k-l} C_l, \qquad k = 0, 1, 2, \cdots. \tag{7}$$

The preceding (7) is an infinite system of recurrence formulas from which all C_k can be determined recursively. Evidently, from (4) and the initial condition in (3),

$$X(0) = C_0 = X_0.$$

Then, from (7),

$$\begin{aligned} C_1 &= A_0 C_0 = A_0 X_0, \\ C_2 &= \tfrac{1}{2}(A_0 C_1 + A_1 C_0) = \tfrac{1}{2}(A_0^2 + A_1)\, X_0, \\ C_3 &= \tfrac{1}{3}(A_0 C_2 + A_1 C_1 + A_2 C_0) \\ &= (\tfrac{1}{6} A_0^3 + \tfrac{1}{6} A_0 A_1 + \tfrac{1}{3} A_1 A_0 + \tfrac{1}{3} A_2)\, X_0. \end{aligned}$$

EXAMPLE. Solve

$$\begin{aligned} y'' - zy' + \gamma y &= 0, \\ y(0) &= 1, \\ y'(0) &= 0. \end{aligned}$$

This differential equation is known as Hermite's differential equation. It arises in many physical problems, in particular in the study of a harmonic oscillator, when studied from the quantum mechanical viewpoint.

Solution. First, by letting

$$\begin{aligned} x_1 &= y, \\ x_2 &= y', \end{aligned}$$

one can rewrite the equation as a system

$$X' = \begin{pmatrix} 0 & 1 \\ -\gamma & z \end{pmatrix} X = (A_0 + zA_1)\, X,$$

$$X(0) = \begin{pmatrix} 1 \\ 0 \end{pmatrix},$$

where

$$A_0 = \begin{pmatrix} 0 & 1 \\ -\gamma & 0 \end{pmatrix},$$

$$A_1 = \begin{pmatrix} 0 & 0 \\ 0 & 1 \end{pmatrix}.$$

We now assume a solution of the form

$$X = \sum_{k=0}^{\infty} C_k z^k$$

and obtain from the differential equation

$$\sum_{k=0}^{\infty} kC_k z^{k-1} = A_0 C_0 + \sum_{k=1}^{\infty} z^k (A_0 C_k + A_1 C_{k-1}).$$

By equating the coefficients of z^k, one obtains

$$C_1 = A_0 C_0,$$
$$C_2 = \tfrac{1}{2}(A_0 C_1 + A_1 C_0),$$
$$C_3 = \tfrac{1}{3}(A_0 C_2 + A_1 C_1),$$
$$\vdots$$
$$C_{k+1} = \frac{1}{k+1}(A_0 C_k + A_1 C_{k-1}),$$

corresponding to (7) in the general development. Evidently

$$C_0 = \begin{pmatrix} 1 \\ 0 \end{pmatrix}$$

from the stipulated initial condition, and then one can easily calculate

$$C_1 = \begin{pmatrix} 0 \\ -\gamma \end{pmatrix}, \qquad C_2 = \begin{pmatrix} -\gamma/2 \\ 0 \end{pmatrix}$$

$$C_3 = \begin{pmatrix} 0 \\ \dfrac{\gamma(\gamma-2)}{3!} \end{pmatrix}, \qquad C_4 = \begin{pmatrix} \dfrac{\gamma(\gamma-2)}{4!} \\ 0 \end{pmatrix}$$

$$C_5 = \begin{pmatrix} 0 \\ -\dfrac{\gamma(\gamma-2)(\gamma-4)}{5!} \end{pmatrix}, \qquad C_6 = \begin{pmatrix} -\dfrac{\gamma(\gamma-2)(\gamma-4)}{6!} \\ 0 \end{pmatrix}$$

One can deduce from these that the general formulas must have the form

$$C_{2k+1} = \begin{pmatrix} 0 \\ \dfrac{(-1)^{k+1}\,\gamma(\gamma-2)(\gamma-4)\cdots(\gamma-2k)}{(2k+1)!} \end{pmatrix},$$

$$C_{2k} = \begin{pmatrix} \dfrac{(-1)^{k}\,\gamma(\gamma-2)(\gamma-4)\cdots(\gamma-2k+2)}{(2k)!} \\ 0 \end{pmatrix},$$

which can now, of course, be proved by mathematical induction.

It is evident that if and only if γ is a positive even integer (say, $2n$), all coefficients beyond a certain index must vanish; that is,

$$C_{2n+1} = C_{2n+2} = C_{2n+3} = \cdots = 0.$$

In this case the solution has only a finite series; that is, it is a polynomial. A similar result is obtained if the initial condition is taken as

$$X_0 = \begin{pmatrix} 0 \\ 1 \end{pmatrix}$$

and γ is a positive odd integer. The resultant polynomials are called Hermite polynomials.

A question that arises in this context naturally, but which up to now has been avoided, is whether the series solution assumed in (4) converges. In other words, will the solutions C_k of the recurence formulas (7) be such that (4) is a convergent series?

Before proceeding to the proof, a useful lemma will be presented, which will later be utilized in the proof.

LEMMA. Let

$$u_r = \frac{K(K+1)\cdots(K+r-1)}{r!},$$

where K is an arbitrary positive number. Then

$$(r+1)\,u_{r+1} = K\sum_{l=0}^{r} u_l.$$

Proof. From the binomial theorem one sees immediately that

$$(1-t)^{-K} = \sum_{r=0}^{\infty} u_r t^r, \qquad |t| < 1.$$

Next, one differentiates with respect to t and obtains

$$K(1 - t)^{-K-1} = \sum_{r=0}^{\infty} ru_r t^{r-1}.$$

But

$$K(1 - t)^{-K-1} = \frac{K}{1 - t}(1 - t)^{-K}$$

$$= K \sum_{r=0}^{\infty} t^r \sum_{r=0}^{\infty} u_r t^r$$

$$= K \sum_{r=0}^{\infty} t^r \sum_{l=0}^{r} u_l,$$

so that

$$\sum_{r=0}^{\infty} ru_r t^{r-1} = K \sum_{r=0}^{\infty} t^r \sum_{l=0}^{r} u_l.$$

By equating the coefficients of t^r, one arrives at the identity to be proved.

We are now ready to state and prove the following theorem.

THEOREM: The differential equation

$$X' = A(z)\,X,$$
$$X(0) = X_0,$$

where $A(z)$ can be represented by

$$A(z) = \sum_{k=0}^{\infty} A_k z^k, \qquad \text{for } |z| < R,$$

has a solution that can be represented by

$$X = \sum_{k=0}^{\infty} C_k z^k,$$

which will converge for $|z| < R$. This solution is uniquely determined by the initial vector X_0.

Proof. In (7) it was shown that

$$(k + 1)\,C_{k+1} = \sum_{l=0}^{k} A_{k-l} C_l, \qquad k = 0, 1, 2, \cdots.$$

One can conclude immediately that

$$(k+1)\,\|C_{k+1}\| \leqslant \sum_{l=0}^{k} \|A_{k-l}\|\,\|C_l\|. \tag{9}$$

If the series for $A(z)$ converges for $|z| < R$, then it will converge absolutely and uniformly for $|z| < p < R$, in which case the terms $\|A_k p^k\|$ must be uniformly bounded. That is, there must exist a constant M such that

$$\|A_k p^k\| \leqslant M, \qquad k = 0, 1, 2, \cdots,$$

so that $\|A_k\| \leqslant M/p^k$. Then it follows from (9) that

$$(k+1)\,\|C_{k+1}\| \leqslant \sum_{l=0}^{k} \frac{\|C_l\|}{p^{k-l}}\,M,$$

and letting

$$p^l\,\|C_l\| = V_l,$$

one can rewrite the last inequality in the form

$$(k+1)\,V_{k+1} \leqslant Mp \sum_{l=0}^{k} V_l. \tag{10}$$

Evidently

$$\begin{aligned}
V_1 &\leqslant MpV_0,\\
V_2 &\leqslant \tfrac{1}{2}\,Mp(V_0 + V_1)\\
&\leqslant \tfrac{1}{2}\,(Mp + M^2p^2)\,V_0\\
&= \frac{Mp(Mp+1)}{2}\,V_0,\\
V_3 &\leqslant \tfrac{1}{3}\,Mp(V_0 + V_1 + V_2)\\
&\leqslant \tfrac{1}{3}\,Mp\left(1 + Mp + \frac{Mp(Mp+1)}{2}\right)V_0\\
&= \frac{Mp(Mp+1)(Mp+2)\,V_0}{3!},
\end{aligned}$$

and one is tempted to deduce that in general one must have

$$V_k \leqslant \frac{Mp(Mp+1)(Mp+2)\cdots(Mp+k-1)}{k!}\,V_0.$$

By assuming the result true for all $r < k$, and referring to the previously proved lemma, one obtains from (10):

$$(r+1)\,V_{r+1} \leqslant MpV_0 \sum_{l=0}^{r} \frac{Mp(Mp+1)\cdots(Mp+l-1)}{l!}$$

$$= (r+1)\frac{Mp(Mp+1)\cdots(Mp+r)}{(r+1)!}\,V_0,$$

so that

$$V_{r+1} \leqslant \frac{Mp(Mp+1)\cdots(Mp+r)}{(r+1)!}\,V_0,$$

which completes the proof by induction. It follows therefore that

$$\| C_k \| \leqslant \frac{Mp(Mp+1)\cdots(Mp+k-1)}{k!p^k}\,\| C_0 \|$$

so that

$$\| X \| = \left\| \sum C_k z^k \right\| \leqslant \| C_0 \| \sum_{k=0}^{\infty} \frac{Mp(Mp+1)\cdots(Mp+k-1)}{k!}\left(\frac{|z|}{p}\right)^k$$

$$= \frac{\| C_0 \|}{\left(1 - \frac{|z|}{p}\right)^{Mp}} \tag{11}$$

Hence the series converges for all $|z| \leqslant p$, and since p can be any positive number less than R, the series converges for all $|z| < R$. From this one can also conclude that in any domain where $A(z)$ has no singularities, X is also free of singularities.

In order to complete the proof, we must show that the solution is unique. Suppose X_1 and X_2 were two different solutions with the same initial vector; then

$$Y = X_1 - X_2$$

is also a solution with initial value.

$$Y(0) = 0.$$

In this case $C_0 = 0$, and it follows from (11) that

$$\| Y \| = 0,$$

in which case $Y \equiv 0$. It also follows from (11) that X depends continuously on X_0.

The expression on the right side of inequality (11) is known as a majorant. This type of proof was first used by Cauchy and has applications not only in the theory of ordinary differential equations, but also in partial differential equations.

The results of this section are evidently existence and uniqueness proofs. Such proofs were furnished in Section 2.7. There however, only the continuity of $A(z)$ was assumed, whereas here the stronger assumption that $A(z)$ be analytic is made. But then again the results are stronger. Whereas the previous proof was not really constructive, this one shows that a solution in the form of an infinite series can be found. Furthermore it was evident that the solution X could have singularities only at those points where $A(z)$ had singularities. The preceding theorem furnished less information regarding the nature of the solution.

3.2 EULER'S DIFFERENTIAL EQUATIONS

The equation

$$X' = \frac{1}{z} AX, \tag{1}$$

where A is a constant matrix, is known as Euler's differential equation. It can be shown that, by a change of independent variable, (1) can be reduced to an equation with a constant coefficient matrix. Let

$$z = e^t$$

so that

$$\frac{d}{dz} X = \frac{d}{dt} X \frac{dt}{dz} = \frac{1}{z} \frac{d}{dt} X,$$

and (1) becomes

$$\frac{d}{dt} X = AX. \tag{2}$$

This equation can now be treated by the methods of Chapter 2, but it will be advantageous to furnish a direct method for future reference. Since (2) can be expected to have solutions of the form

$$X = e^{\lambda t} C,$$

one can seek solutions of (1) in the form

$$X = z^{\lambda} C. \tag{3}$$

Insertion of (3) in (1) leads to

$$\lambda z^{\lambda - 1} C = \frac{1}{z} A z^{\lambda} C,$$

which can be rearranged to yield

$$z^{\lambda-1}(A - \lambda I)\,C = 0. \tag{4}$$

It follows from (4) that if

$$|\,A - \lambda I\,| = 0,$$

that is, if λ is an eigenvalue of A, then it will be possible to find an eigenvector C such that

$$(A - \lambda I)\,C = 0.$$

Once these are found, a solution (3) can be constructed.

EXAMPLE 1. Solve

$$z^2y'' - 2zy' + 2y = 0.$$

Solution. Let

$$x_1 = y$$

$$x_2 = zy'$$

$$x_1' = y' = \frac{1}{z}\,x_2$$

$$x_2' = zy'' + y' = \left(2y' - \frac{2}{z}\,y\right) + y' = 3y' - \frac{2}{z}\,y = \frac{3}{z}\,x_2 - \frac{2}{z}\,x_1.$$

This system can now be rewritten as

$$X' = \frac{1}{z}\begin{pmatrix} 0 & 1 \\ -2 & 3 \end{pmatrix} X,$$

which is an Euler equation. To find a solution in the form

$$X = z^{\lambda}C,$$

we first seek the eigenvalues of the matrix:

$$|\,A - \lambda I\,| = \begin{vmatrix} -\lambda & 1 \\ -2 & 3 - \lambda \end{vmatrix} = \lambda^2 - 3\lambda + 2 = 0,$$

$$\lambda_1 = 1, \qquad \lambda_2 = 2.$$

The corresponding eigenvector are

$$C_1 = \begin{pmatrix} a \\ a \end{pmatrix}, \qquad C_2 = \begin{pmatrix} b \\ 2b \end{pmatrix}.$$

The complete solution now becomes

$$X = z\begin{pmatrix} a \\ a \end{pmatrix} + z^2\begin{pmatrix} b \\ 2b \end{pmatrix}.$$

and $y = x_1 = az + bz^2$.

One can show that in general the equation

$$z^n y^{(n)} + a_1 z^{n-1} y^{(n-1)} + a_2 z^{n-2} y^{(n-2)} + \cdots + a_{n-1} z y' + a_n y = 0$$

can be reduced to the form (1) by the substitutions

$$\begin{aligned} x_1 &= y, \\ x_2 &= zy', \\ x_3 &= z^2 y'', \\ &\vdots \\ x_n &= z^{n-1} y^{(n-1)}. \end{aligned}$$

For then

$$\begin{aligned} x_1' &= \frac{1}{z} x_2, \\ x_2' &= \frac{1}{z}(x_2 + x_3), \\ x_3' &= \frac{1}{z}(2x_3 + x_4), \\ &\vdots \\ x_k' &= \frac{1}{z}((k-1)\, x_k + x_{k+1}) \\ &\vdots \\ x_n' &= \frac{1}{z}((n - 1 - a_1)\, x_n - a_2 x_{n-2} \cdots - a_{n-1} x_2 - a_n x_1), \end{aligned}$$

and when written in matrix form,

$$X' = \frac{1}{z}\begin{pmatrix} 0 & 1 & 0 & 0 & 0 & \cdots & 0 & 0 \\ 0 & 1 & 1 & 0 & 0 & \cdots & 0 & 0 \\ 0 & 0 & 2 & 1 & 0 & \cdots & 0 & 0 \\ \vdots & \vdots & \vdots & \vdots & \vdots & \cdots & \vdots & \vdots \\ 0 & 0 & 0 & 0 & 0 & \cdots & n-2 & 1 \\ -a_n & -a_{n-1} & -a_{n-2} & -a_{n-3} & -a_{n-4} & \cdots & -a_2 & n-1-a_1 \end{pmatrix} X$$

It is evident that as long as A has distinct eigenvalues, one can find n linearly independent solutions of type (3). But in the case of multiple eigenvalues one has to resort to the methods of Chapter 2 in the corresponding case.

EXAMPLE 2. Solve

$$z^3y''' + zy' - y = 0.$$

Solution. Let

$$x_1 = y$$
$$x_2 = zy'$$
$$x_3 = z^2y''$$

Then

$$x_1' = \frac{1}{z}x_2,$$
$$x_2' = \frac{1}{z}(x_2 + x_3),$$
$$x_3' = \frac{1}{z}(2x_3 - x_2 + x_1),$$

so that

$$X' = \frac{1}{z}\begin{pmatrix} 0 & 1 & 0 \\ 0 & 1 & 1 \\ 1 & -1 & 2 \end{pmatrix} X.$$

One can check that

$$|A - \lambda I| = \begin{vmatrix} -\lambda & 1 & 0 \\ 0 & 1-\lambda & 1 \\ 1 & -1 & 2-\lambda \end{vmatrix} = -(\lambda - 1)^3,$$

so that $\lambda = 1$ is a triple eigenvalue. One solution is seen to be given by

$$X_1 = z\begin{pmatrix} a \\ a \\ 0 \end{pmatrix}.$$

To find other solutions, we let

$$X = \begin{pmatrix} 1 & az & 0 \\ 0 & az & 0 \\ 0 & 0 & 1 \end{pmatrix} Y$$

in accordance with the method of Section 2.5. Then one finds that

$$Y' = \begin{pmatrix} 0 & 0 & -\frac{1}{z} \\ 0 & 0 & \frac{1}{az^2} \\ \frac{1}{z} & 0 & \frac{2}{z} \end{pmatrix} Y. \tag{5}$$

The order of this system can now be reduced by writing

$$\begin{pmatrix} y_1 \\ y_3 \end{pmatrix}' \equiv \bar{Y}' = \frac{1}{z}\begin{pmatrix} 0 & -1 \\ 1 & 2 \end{pmatrix} \bar{Y}. \tag{6}$$

This is an Euler equation, and the eigenvalues are given by

$$| A - \lambda I | = \begin{vmatrix} -\lambda & -1 \\ 1 & 2-\lambda \end{vmatrix} = (\lambda - 1)^2.$$

It follows that one solution is given by

$$\bar{Y} = z\begin{pmatrix} b \\ -b \end{pmatrix}.$$

Returning to (5) we have

$$y_2' = \frac{1}{az^2}\, y_3 = \frac{-b}{az},$$

so that

$$y_2 = -\frac{b}{a}\ln z + c.$$

It follows that

$$X_2 = \begin{pmatrix} 1 & az & 0 \\ 0 & az & 0 \\ 0 & 0 & 1 \end{pmatrix}\begin{pmatrix} bz \\ \dfrac{-b}{a}\ln z + c \\ -bz \end{pmatrix} = \begin{pmatrix} bz - bz\ln z + acz \\ -bz\ln z + acz \\ -bz \end{pmatrix}$$

Apparently c is a third arbitrary parameter, but the terms in X_2 containing c already appear in X_1. Therefore we can choose $c = 0$ without loss of generality. To find a third solution, we return to (6) and let

$$\bar{Y} = \begin{pmatrix} 1 & bz \\ 0 & -bz \end{pmatrix} \bar{\bar{Y}},$$

and one finds that

$$\bar{\bar{Y}}' = \frac{1}{z}\begin{pmatrix} 1 & 0 \\ \dfrac{-1}{bz} & 0 \end{pmatrix} \bar{\bar{Y}}.$$

It follows that

$$\bar{\bar{Y}} = \begin{pmatrix} cz \\ -\dfrac{c}{b}\ln z \end{pmatrix},$$

and then

$$\bar{Y} = \begin{pmatrix} cz - cz\ln z \\ cz\ln z \end{pmatrix}.$$

By returning to (5), we obtain

$$y_2' = \frac{c}{az} \ln z,$$

so that

$$y_2 = \frac{c}{2a} (\ln z)^2.$$

Finally

$$X = \begin{pmatrix} 1 & az & 0 \\ 0 & az & 0 \\ 0 & 0 & 1 \end{pmatrix} \begin{pmatrix} cz - cz \ln z \\ \frac{c}{2a} (\ln z)^2 \\ cz \ln z \end{pmatrix} = \begin{pmatrix} cz - cz \ln z + \frac{cz}{2} (\ln z)^2 \\ \frac{c}{2} z(\ln z)^2 \\ cz \ln z \end{pmatrix}.$$

The complete solution now is

$$X = \begin{pmatrix} az \\ az \\ 0 \end{pmatrix} + \begin{pmatrix} bz - bz \ln z \\ -bz \ln z \\ -bz \end{pmatrix} + \begin{pmatrix} cz - cz \ln z + \frac{cz}{2} (\ln z)^2 \\ \frac{c}{2} z(\ln z)^2 \\ cz \ln z \end{pmatrix},$$

and finally,

$$y = x_1 = (a + b + c)\, z - (b + c)\, z \ln z + \frac{c}{2} z(\ln z)^2.$$

3.3 EQUATIONS WITH REGULAR SINGULARITIES

Differential equations of the type

$$X' = \frac{1}{z - z_0} A(z)\, X, \tag{1}$$

where $A(z)$ is an analytic function near z_0, are said to have regular, or Fuchsian, singularities at z_0. The latter name is in honor of the mathematician Fuchs, who contributed to the theory underlying these equations. Without loss of generality one can introduce a shift in coordinates so as to place the singularity at the origin, in which case one can write

$$X' = \frac{1}{z} \sum_{k=0}^{\infty} A_k z^k X. \tag{2}$$

The equation

$$y^{(n)} + \frac{a_1(z)}{z} y^{(n-1)} + \frac{a_2(z)}{z^2} y^{(n-2)} + \cdots + \frac{a_{n-1}(z)}{z^{n-1}} y' + \frac{a_n(z)}{z^n} y = 0 \tag{3}$$

where all $a_i(z)$ are analytic functions near $z = 0$, can be reduced to the form (2) by the substitutions

$$\begin{aligned} x_1 &= y, \\ x_2 &= zy', \\ x_3 &= z^2y'' \\ &\vdots \\ x_n &= z^{n-1}y^{(n-1)}. \end{aligned}$$

Then one obtains the system

$$X' = \frac{1}{z}\begin{vmatrix} 0 & 1 & 0 & 0 & 0 & 0 & 0 \\ 0 & 1 & 1 & 0 & 0 & 0 & 0 \\ 0 & 0 & 2 & 1 & 0 & 0 & 0 \\ \vdots & \vdots & \vdots & \vdots & \vdots & \vdots & \vdots \\ 0 & 0 & 0 & 0 & 0 & n-2 & 1 \\ -a_n(z) & -a_{n-1}(z) & -a_{n-2}(z) & a_{n-3}(z) & a_{n-4}(z) & -a_2(z) & n-1-a_1(z) \end{vmatrix} X$$

equivalent to (3). This is similar to the comparable step in the preceding section except that there all a_i were constants.

EXAMPLE 1. The equation

$$y'' + \frac{1}{z}y' + \frac{z^2 - n^2}{z^2}y = 0$$

is known as Bessel's equation. It arises in many problems of mathematical physics, in particular in heat flow and wave propagation problems in cylindrical and spherical coordinate systems.

Solution. By using the preceding substitutions, one can reduce this equation to the system

$$X' = \frac{1}{z}\begin{pmatrix} 0 & 1 \\ n^2 - z^2 & 0 \end{pmatrix} X = \frac{1}{z}(A_0 + A_2z^2)\, X, \tag{4}$$

where

$$A_0 = \begin{pmatrix} 0 & 1 \\ n^2 & 0 \end{pmatrix}, \qquad A_2 = \begin{pmatrix} 0 & 0 \\ -1 & 0 \end{pmatrix}.$$

In Section 3.1 it was shown that if the matrix coefficient of the equation is analytic, then the solution is also analytic. This does not preclude the possibility of the solution being analytic even when the matrix coefficient is not. We shall therefore attempt to find a solution of (4) that is analytic and therefore can be expressed in the form

$$X = \sum_{k=0}^{\infty} C_k z^k. \tag{5}$$

Insertion of this trial solution in (4) leads to

$$\sum_{k=0}^{\infty} kC_k z^{k-1} = \sum_{k=0}^{\infty} A_0 C_k z^{k-1} + \sum_{k=0}^{\infty} A_2 C_k z^{k+1}.$$

Comparison of coefficients of like powers of z leads to the recurrence equations

$$\begin{aligned} A_0 C_0 &= 0, \\ (A_0 - I)\, C_1 &= 0, \\ (A_0 - 2I)\, C_2 &= -A_2 C_0, \\ (A_0 - 3I)\, C_3 &= -A_2 C_1, \\ &\vdots \\ (A_0 - kI)\, C_k &= -A_2 C_{k-2}. \end{aligned} \tag{6}$$

It follows from these equations that a nonvanishing C_0 can be found only if A_0 has an eigenvalue $\lambda = 0$. The eigenvalues of A_0 are given by

$$\begin{vmatrix} -\lambda & 1 \\ n^2 & -\lambda \end{vmatrix} = \lambda^2 - n^2 = 0,$$

that is, $\lambda = \pm n$. It follows that the parameter n must vanish in order to obtain a type (5) solution with nonvanishing C_0. One finds in this case that

$$C_0 = \begin{pmatrix} a \\ 0 \end{pmatrix}, \qquad C_1 = \begin{pmatrix} 0 \\ 0 \end{pmatrix},$$

and that all C_k with odd subscripts must vanish follows from (6). For the C_k with even subscripts one obtains

$$\begin{pmatrix} -k & 1 \\ 0 & -k \end{pmatrix} C_k = \begin{pmatrix} 0 & 0 \\ 1 & 0 \end{pmatrix} C_{k-2},$$

which can be solved for C_k to yield

$$C_k = \begin{pmatrix} -\dfrac{1}{k^2} & 0 \\ -\dfrac{1}{k} & 0 \end{pmatrix} C_{k-2}.$$

One can now verify by induction that

$$C_{2k} = \begin{pmatrix} \dfrac{(-1)^k a}{[2 \cdot 4 \cdot 6 \cdots (2k)]^2} \\ \dfrac{(-1)^k a(2k)}{[2 \cdot 4 \cdot 6 \cdots (2k)]^2} \end{pmatrix} = \frac{(-1)^k a}{4^k (k!)^2} \begin{pmatrix} 1 \\ 2k \end{pmatrix} \tag{7}$$

Then a solution of

$$X' = \frac{1}{z} \begin{pmatrix} 0 & 1 \\ -z^2 & 0 \end{pmatrix} X$$

is given by

$$X = \sum_{k=0}^{\infty} C_{2k} z^{2k}$$

and the C_{2k} are given in (7).

Returning to (6) we see that if $n = 1$, it would follow that $C_0 = 0$, but now C_1 would be obtained from the second equation and all subsequent C_k could be found. More generally, if n is a positive integer, one can conclude from (6) that

$$\begin{aligned} C_0 = C_1 = \cdots = C_{n-1} &= 0, \\ (A_0 - nI)\, C_n &= 0, \\ (A_0 - (n+1)\, I)\, C_{n+1} &= 0, \\ (A_0 - (n+2)\, I)\, C_{n+2} &= -A_2 C_n \\ &\vdots \\ (A_0 - kI)\, C_k &= -A_2 C_{k-2}. \end{aligned}$$

One can show very easily now that

$$C_{n+2k} = \frac{(-1)^k\, an!}{4^k k! (n+k)!} \begin{pmatrix} 1 \\ 2k+n \end{pmatrix}, \qquad k = 0, 1, 2, \cdots, \tag{8}$$

so that a solution of (4), provided n is a positive integer, is given by

$$X = \sum_{k=0}^{\infty} C_{n+2k}\, z^{n+2k},$$

where the coefficients are defined in (8). A second linearly independent solution could be obtained by using this solution to reduce the order of the system.

This example clearly illustrates that an analytic solution of (2) can be obtained if the matrix A_0 has an eigenvalue that is a positive integer. If no such eigenvalue exists, it will not be possible to find an analytic solution. But it will now be shown that (2) can be transformed into a new equation of type (2), where the new leading matrix coefficient A_0 will have at least one eigenvalue $\lambda = 0$.

To do so, we return to (2) and assume that μ is an eigenvalue of A_0. Then we introduce a new dependent variable:

$$X = z^{\mu} Y,$$

which leads to

$$\begin{aligned} Y' &= \frac{1}{z} [A(z) - \mu I] Y \\ &= \frac{1}{z} \Big[\sum_{k=0}^{\infty} A_k z^k - \mu I \Big] Y. \end{aligned} \tag{9}$$

If one now assumes that Y can be expressed in the form

$$Y = \sum_{k=0}^{\infty} C_k z^k,$$

one obtains the set of recurrence equations

$$\begin{aligned} (A_0 - \mu I) C_0 &= 0, \\ (A_0 - (\mu + 1) I) C_1 &= -A_1 C_0, \\ (A_0 - (\mu + k) I) C_k &= -\sum_{l=0}^{k-1} A_{k-l} C_l. \end{aligned} \tag{10}$$

These steps should be compared to the steps preceding (7) in Section 3.1. One important difference is that in Section 3.1, C_0 could be prescribed arbitrarily and then the succeeding C_k could be uniquely determined. Here C_0 is determined from the first of the equations in (10). That a nonvanishing value of C_0 can be found follows from the fact that μ is an eigenvalue of A_0, so that $| A_0 - \mu I | = 0$. Then $C_1, C_2 \cdots$ can be determined from the remaining equations in (10). However, the process may fail if $| A_0 - (\mu + k) I | = 0$ for some positive integer k. It need not, if the corresponding equation is compatible.

We shall now show that when $| A_0 - (\mu + k) I | \neq 0$ for all $k \geqslant 1$, then the series for X, namely,

$$X = z^{\mu} \sum_{k=0}^{\infty} C_k z^k \tag{11}$$

will converge. To do so, we observe, as in Section 3.1, that for some p and M we must have

$$\| A_k p^k \| \leqslant M, \qquad k = 0, 1, 2, \cdots.$$

From the assumption that $|A_0 - (\mu + k)I| \neq 0$ for all positive integers k, and since

$$\lim_{k\to\infty} \frac{\| A_0 - (\mu + k)I \|}{k} = n,$$

we see that there must exist a positive quantity ϵ such that

$$\frac{\| A_0 - (\mu + k)I \|}{k} > \epsilon, \qquad k = 1, 2, \cdots.$$

Then the typical equation in (10) can be written as

$$(k+1)\,C_{k+1} = -[A_0 - (\mu + k + 1)I]^{-1}\,(k+1) \quad \sum_{l=0}^{k} A_{k+1-l}C_l.$$

It now follows, as in Section 3.1, that

$$(k+1)\,\| C_{k+1} \|\, p^{k+1} < \frac{M}{\epsilon} \sum_{l=0}^{k} \| C_l \|\, p^l,$$

from which one deduces as before that

$$\| C_k \| \leqslant \frac{(M/\epsilon)\,(M/\epsilon + 1)\cdots(M/\epsilon + k - 1)}{k!\,p^k} \| C_0 \|.$$

By applying these results to (11), one obtains

$$\| X \| = \| z^\mu \sum_{k=0}^{\infty} C_k z^k \| \leqslant | z^\mu | \sum_{k=0}^{\infty} \| C_k \|\, | z |^k$$

$$\leqslant | z^\mu |\, \| C_0 \| \sum_{k=0}^{\infty} \frac{M/\epsilon(M/\epsilon + 1)\cdots(M/\epsilon + k - 1)}{k!} \left(\frac{| z |}{p}\right)^k$$

$$= \frac{| z^\mu |\, \| C_0 \|}{\left(1 - \frac{| z |}{p}\right)^{M/\epsilon}}.$$

From this expression one deduces immediately that (11) converges wherever the series

$$\sum_{k=0}^{\infty} A_k z^k$$

converges and that X depends continuously on C_0.

The preceding results will now be combined into the following theorem.

THEOREM: The differential equation

$$X' = \frac{1}{z}\Big(\sum_{k=0}^{\infty} A_k z^k\Big) X,$$

where $\sum_{k=0}^{\infty} A_k z^k$ converges for $|z| < R$ will have solutions of the form

$$X = z^{\lambda} \sum_{k=0}^{\infty} C_k z^k$$

which also converge for $|z| < R$, provided λ is an eigenvalue of A_0 and no other eigenvalue of the form $\lambda + k$, k a positive integer, exists.

We shall now consider some of the special cases that can arise when two eigenvalues of the form λ and $\lambda + k$ occur.

EXAMPLE 2. Find the general solution for the equation

$$X' = \begin{pmatrix} 0 & \frac{1}{z} \\ -z & 0 \end{pmatrix} X. \tag{13}$$

Solution. This is the problem treated in Example 1, with $n = 0$. There it was seen that a solution is

$$X_1 = \sum_{k=0}^{\infty} \frac{(-1)^k z^{2k}}{4^k (k!)^2} \binom{1}{2k}.$$

This solution has two components

$$x_1 = \sum_{k=0}^{\infty} \frac{(-1)^k z^{2k}}{4^k (k!)^2},$$

$$x_2 = \sum_{k=0}^{\infty} \frac{(-1)^k 2k z^{2k}}{4^k (k!)^2}. \tag{14}$$

One easily sees, either by a direct calculation or from (13), that

$$x_1' = \frac{1}{z} x_2$$

$$x_2' = -z x_1,$$

which will be useful in the sequel. Next let

$$X = \begin{pmatrix} 1 & x_1 \\ 0 & x_2 \end{pmatrix} Y$$

so that, from (13),

$$\begin{pmatrix} 0 & \frac{1}{z} x_2 \\ 0 & -zx_1 \end{pmatrix} Y + \begin{pmatrix} 1 & x_1 \\ 0 & x_2 \end{pmatrix} Y' = \begin{pmatrix} 0 & \frac{1}{z} x_2 \\ -z & -zx_1 \end{pmatrix} Y,$$

and solving for Y', we obtain

$$Y' = \begin{pmatrix} z\dfrac{x_1}{x_2} & 0 \\ -\dfrac{z}{x_2} & 0 \end{pmatrix} Y.$$

Then we find that

$$y_1' = z\frac{x_1}{x_2} y_1 = -\frac{x_2'}{x_2} y_1,$$

$$y_2' = -\frac{z}{x_2} y_1,$$

so that

$$y_1 = \frac{c_1}{x_2},$$

$$y_2 = -c_1 \int \frac{t}{x_2^2(t)}\, dt.$$

It follows that we can find a new solution X_2 given by

$$X_2 = \begin{pmatrix} \dfrac{c_1}{x_2} - c_1 x_1 \displaystyle\int \frac{t}{x_2^2(t)}\, dt \\ -c_1 x_2 \displaystyle\int \frac{t}{x_2^2(t)}\, dt \end{pmatrix}$$

The first component of this solution can be simplified as follows:

$$\begin{aligned} \frac{1}{x_2} - x_1 \int \frac{t}{x_2^2}\, dt &= \frac{1}{x_2} - x_1 \int \frac{-(x_2'/x_1)}{x_2^2}\, dt \\ &= -x_1 \int \frac{x_1'}{x_2 x_1^2}\, dt \\ &= -x_1 \int \frac{1}{t x_1^2}\, dt, \end{aligned}$$

so that

$$X_2 = -c_1 \begin{pmatrix} x_1 \displaystyle\int^z \frac{1}{t x_1^2}\, dt \\ x_2 \displaystyle\int^z \frac{t}{x_2^2}\, dt \end{pmatrix}.$$

Since, from (14),

$$x_1 = 1 - \frac{z^2}{4} + \cdots,$$

$$x_2 = -\frac{z^2}{2} + \frac{z^4}{16} + \cdots,$$

we see that

$$x_1 \int^z \frac{1}{tx_1^2}\,dt = x_1 \int^z \frac{1 + \frac{t^2}{2} + \cdots}{t}\,dt$$

$$= x_1 \ln z + x_1\left(\frac{z^2}{4} + \cdots\right)$$

$$x_2 \int^z \frac{t}{x_2^2}\,dt = 4x_2 \int^z \frac{t\left(1 + \frac{t^2}{4} + \cdots\right)}{t^4}\,dt$$

$$= \frac{-2x_2}{z^2} + x_2 \ln z + \cdots$$

It follows that

$$X_2 = c_1 \ln z \begin{pmatrix} x_1 \\ x_2 \end{pmatrix} + \cdots,$$

showing that, as a consequence of the double eigenvalue at $\lambda = 0$, logarithmic terms have entered.

The following example shows that such logarithmic terms need not be introduced.

EXAMPLE 3. Solve

$$X' = \left[\frac{1}{z}\begin{pmatrix} 0 & 1 \\ 0 & 2 \end{pmatrix} + \begin{pmatrix} 0 & 0 \\ -1 & 1 \end{pmatrix}\right] X.$$

Solution. The leading matrix coefficient

$$\begin{pmatrix} 0 & 1 \\ 0 & 2 \end{pmatrix}$$

has two eigenvalues, $\lambda_1 = 0$, and $\lambda_2 = 2$. According to the general theory we must therefore be able to obtain a solution of the form

$$X = z^2 \sum_{k=0}^{\infty} C_k z^k.$$

But we are not certain that we can obtain a solution corresponding to the smaller eigenvalue, since their difference is a positive integer. Nevertheless we shall try to find a solution of the form

$$X = \sum_{k=0}^{\infty} C_k z^k.$$

Then it follows that the C_k must satisfy the recurrence formulas:

$$\begin{pmatrix} k & -1 \\ 0 & k-2 \end{pmatrix} C_k = \begin{pmatrix} 0 & 0 \\ -1 & 1 \end{pmatrix} C_{k-1}, \qquad k = 1, 2, \cdots. \tag{15}$$

C_0 must satisfy

$$\begin{pmatrix} 0 & -1 \\ 0 & -2 \end{pmatrix} C_0 = 0,$$

so that

$$C_0 = \begin{pmatrix} a \\ 0 \end{pmatrix}.$$

C_1 is found from

$$\begin{pmatrix} 1 & -1 \\ 0 & -1 \end{pmatrix} C_1 = \begin{pmatrix} 0 & 0 \\ -1 & 1 \end{pmatrix} C_0 = \begin{pmatrix} 0 \\ -a \end{pmatrix},$$

so that

$$C_1 = \begin{pmatrix} a \\ a \end{pmatrix}.$$

Next, to find C_2, we examine the equation

$$\begin{pmatrix} 2 & -1 \\ 0 & 0 \end{pmatrix} C_2 = \begin{pmatrix} 0 & 0 \\ -1 & 1 \end{pmatrix} C_1 = \begin{pmatrix} 0 \\ 0 \end{pmatrix}.$$

The determinant of this system is 0, but the term on the right is also 0. Hence we do obtain a solution for C_2:

$$C_2 = \begin{pmatrix} b \\ 2b \end{pmatrix}.$$

Now we can solve for all the remaining C_k, since for $k > 2$, (15) is always solvable. As a matter of fact we have in this way obtained the complete solution to our problem because

$$X_1 = \begin{pmatrix} a \\ 0 \end{pmatrix} + \begin{pmatrix} a \\ a \end{pmatrix} z$$

forms one solution and

$$X_2 = \begin{pmatrix} b \\ 2b \end{pmatrix} z^2 + \cdots$$

will be a second solution. The first solution is in this case a polynomial. To find the second solution, it is easier to make use of X_1 rather than try to solve the equations (15). Then let

$$X = \begin{pmatrix} 1 & 1+z \\ 0 & z \end{pmatrix} Y,$$

and one finds that

$$Y' = \begin{pmatrix} 1 + \frac{1}{z} & 0 \\ -\frac{1}{z} & 0 \end{pmatrix} Y.$$

From this one can show that

$$X_2 = -2b \begin{pmatrix} 1 + z - e^z \\ z - ze^z \end{pmatrix}.$$

3.4 ASYMPTOTIC SOLUTIONS

In the previous sections various methods were investigated for finding solutions for finite values of z. Since these were series in positive powers of z, they were of use only for moderately small values of z. Such methods would be of little value in the case of an equation of the form

$$y' = e^{1/z} y. \tag{1}$$

Although one can solve this equation explicitly, the answer cannot be expressed in terms of elementary functions. But if one were interested in the behaviour of y for large values of z, one could proceed as follows: Expand $e^{1/z}$ in a series in terms of $1/z$ so that (1) can be rewritten in the form

$$y' = \left(1 + \frac{1}{z} + \frac{1}{2z^2} + \frac{1}{6z^3} + \cdots\right) y. \tag{2}$$

We shall consider the more general equation

$$y' = \left(a + \frac{b}{z} + \frac{c}{z^2} + \cdots\right) y$$

and write it in the form

$$\frac{dy}{y} = \left(a + \frac{b}{z} + \frac{c}{z^2} + \cdots\right) dz,$$

so that we obtain

$$\ln y = \ln c_0 + az + b \ln z - \frac{c}{z} + \cdots,$$

where $\ln c_0$ is a constant of integration. Solving for y, we have

$$y = c_0 e^{az} z^b e^{-c/z+\cdots} = c_0 e^{az} z^b \left(1 - \frac{c}{z} + \cdots\right). \tag{3}$$

The resultant series can now be expected to yield information about y for very large values of $|z|$.

One can find solutions of similar character in the case of systems of the form

$$X' = \sum_{k=0}^{\infty} A_k z^{-k} X. \tag{4}$$

We shall now show that we can find numbers λ and μ and vectors C_k such that

$$X = e^{\lambda z} z^{\mu} \sum_{k=0}^{\infty} C_k z^{-k} \tag{5}$$

is a solution of (4). As in previous sections we shall insert (5) in (4) and compare coefficients of corresponding powers of z. But we shall make one stipulation, namely, we shall require that A_0 have distinct eigenvalues. The reasons for this will emerge presently, and the case where A_0 has multiple eigenvalues becomes much more complicated. Then

$$\lambda e^{\lambda z} z^{\mu} \sum_{k=0}^{\infty} C_k z^{-k} + \mu e^{\lambda z} z^{\mu} \sum_{k=0}^{\infty} C_k z^{-k-1} - e^{\lambda z} z^{\mu} \sum_{k=0}^{\infty} k C_k z^{-k-1}$$
$$= e^{\lambda z} z^{\mu} \sum_{k=0}^{\infty} z^{-k} \sum_{l=0}^{k} A_{k-l} C_l,$$

so that we obtain

$$\begin{aligned} &(A_0 - \lambda I)\, C_0 = 0, \\ &(A_0 - \lambda I)\, C_1 = \mu C_0 - A_1 C_0, \\ &(A_0 - \lambda I)\, C_k = (\mu - k + 1)\, C_{k-1} - \sum_{l=0}^{k-1} A_{k-l} C_l, \qquad k = 2, 3, \cdots. \end{aligned} \tag{6}$$

One novel aspect of the system of recurrence formulas (6) is the fact that the determinant of all equations must vanish. This follows from the fact that all equations have the same determinant, and in order for a nonvanishing C_0

to exist, it will be necessary that $| A_0 - \lambda I | = 0$. Therefore λ must be an eigenvalue and C_0 an eigenvector. We turn our attention now to the second of equations (6), namely,

$$(A_0 - \lambda I)\, C_1 = \mu C_0 - A_1 C_0.$$

In general such an equation will not have a solution unless the right side satisfies certain compatibility conditions. But this can be achieved by observing that by a suitable choice of μ, which up to now was still undetermined, the system can be made compatible.

It is precisely at this point that the necessity of λ being a simple eigenvalue comes in. If λ were a multiple eigenvalue, then it could turn out that no choice of μ would make this equation compatible. Hence, if we hope to obtain n linearly independent solutions of type (5), we have to require that A_0 have distinct eigenvalues. That is not to say that there will not be n linearly independent solutions otherwise, but they will not be of type (5) necessarily.

We have now seen that C_1 has been determined, but not uniquely. Effectively the determining equation for C_1 becomes a system of $n - 1$ equations in the n unknown components of C_1, and an arbitrary element is introduced. We now turn to the third of equations (6):

$$(A_0 - \lambda I)\, C_2 = (\mu - 1)\, C_1 - A_1 C_1 - A_2 C_0.$$

To make this equation compatible, the arbitrary element in C_1 must be properly selected. Then C_1 becomes fully determined and now C_2 can be determined, but with some arbitrary element in it. By going on to the next stage, the equation for C_3, the necessary compatibility requirement will determine C_2 uniquely. In this fashion all C_k can be determined.

Example 1. Determine the solutions for large values of $| z |$ of the equation

$$y'' + \frac{1}{z} y' + \left(1 - \frac{n^2}{z^2}\right) y = 0.$$

Solution. This is again Bessel's equation, which was earlier examined and for which solutions for relatively small values of $| z |$ were determined. Let

$$x_1 = y,$$
$$x_2 = y',$$

so that

$$X' = \left[\begin{pmatrix} 0 & 1 \\ -1 & 0 \end{pmatrix} + \frac{1}{z}\begin{pmatrix} 0 & 0 \\ 0 & -1 \end{pmatrix} + \frac{1}{z^2}\begin{pmatrix} 0 & 0 \\ n^2 & 0 \end{pmatrix}\right] X.$$

To find a solution of the type

$$X = e^{\lambda z} z^{\mu} \sum_{k=0}^{\infty} C_k z^{-k},$$

we must solve (6). First, to determine C_0, we require that

$$|A_0 - \lambda I| = \begin{vmatrix} -\lambda & 1 \\ -1 & -\lambda \end{vmatrix} = \lambda^2 + 1 = 0,$$

so that $\lambda = \pm i$. The roots are simple and we shall work with $\lambda = i$.

$$\begin{pmatrix} -i & 1 \\ -1 & -i \end{pmatrix} C_0 = 0,$$

so that

$$C_0 = a \begin{pmatrix} i \\ -1 \end{pmatrix}.$$

To determine C_1, we turn to the next equation:

$$\begin{pmatrix} -i & 1 \\ -1 & -i \end{pmatrix} C_1 = a \begin{pmatrix} \mu & 0 \\ 0 & \mu + 1 \end{pmatrix} \begin{pmatrix} i \\ -1 \end{pmatrix}.$$

It will be convenient to multiply both sides of this equation by the nonsingular matrix

$$\begin{pmatrix} 1 & 0 \\ i & 1 \end{pmatrix}. \tag{7}$$

Then the equation becomes

$$\begin{pmatrix} -i & 1 \\ 0 & 0 \end{pmatrix} C_1 = a \begin{pmatrix} \mu & 0 \\ i\mu & \mu + 1 \end{pmatrix} \begin{pmatrix} i \\ -1 \end{pmatrix} = a \begin{pmatrix} i\mu \\ -1 - 2\mu \end{pmatrix}$$

and the evident compatibility requirement is that $\mu = -1/2$. With this choice we find that

$$C_1 = a \begin{pmatrix} b \\ ib - i/2 \end{pmatrix}.$$

Next we turn to

$$\begin{pmatrix} -i & 1 \\ -1 & -i \end{pmatrix} C_2 = a \begin{pmatrix} -3/2 & 0 \\ 0 & -1/2 \end{pmatrix} \begin{pmatrix} b \\ ib - i/2 \end{pmatrix} - a \begin{pmatrix} 0 & 0 \\ n^2 & 0 \end{pmatrix} \begin{pmatrix} i \\ -1 \end{pmatrix},$$

and to simplify, we again multiply by (7):

$$\begin{pmatrix} -i & 1 \\ 0 & 0 \end{pmatrix} C_2 = a \begin{pmatrix} -3/2b \\ -2ib - (n^2 - 1/4)\, i \end{pmatrix}.$$

The compatibility requirements yields b, so that

$$b = -\frac{1}{2}\left(n^2 - \frac{1}{4}\right)$$

and

$$C_2 = a\begin{pmatrix} c \\ ic + \frac{3}{4}\left(n^2 - \frac{1}{4}\right) \end{pmatrix}, \qquad C_1 = a\begin{pmatrix} -\frac{1}{2}\left(n^2 - \frac{1}{4}\right) \\ -\frac{i}{2}n^2 - i\frac{3}{8} \end{pmatrix}.$$

From the next equation one finds that

$$C_2 = a\begin{pmatrix} \frac{i}{8}\left(n^2 - \frac{1}{4}\right)\left(n^2 - \frac{9}{4}\right) \\ -\frac{1}{8}\left(n^2 - \frac{1}{4}\right)\left(n^2 - \frac{25}{4}\right) \end{pmatrix}.$$

Therefore one finds that

$$X_1 = ae^{iz}z^{-1/2}\left[\begin{pmatrix} i \\ -1 \end{pmatrix} + \begin{pmatrix} -\frac{1}{2}\left(n^2 - \frac{1}{4}\right) \\ -\frac{i}{2}\left(n^2 + \frac{3}{4}\right) \end{pmatrix}\frac{1}{z} + \begin{pmatrix} \frac{i}{8}\left(n^2 - \frac{1}{4}\right)\left(n^2 - \frac{9}{4}\right) \\ -\frac{1}{8}\left(n^2 - \frac{1}{4}\right)\left(n^2 - \frac{25}{4}\right) \end{pmatrix}\frac{1}{z^2} + \cdots\right].$$

It is interesting that the leading term, which predominates for large values of z, is independent of n. Furthermore, for $n^2 = 1/4$, the third and all succeeding terms will vanish. In this case an exact solution will have been obtained. One can show that whenever $n^2 = m^2/4$, where m is an odd integer, all but a finite number of terms will vanish, so that an exact solution will be obtained.

Since the differential equation contains only real terms, one can easily construct a second solution merely by taking the conjugate of the first. Then

$$X_2 = be^{-iz}z^{-1/2}\left[\begin{pmatrix} -i \\ -1 \end{pmatrix} + \begin{pmatrix} -\frac{1}{2}\left(n^2 - \frac{1}{4}\right) \\ \frac{i}{2}\left(n^2 + \frac{3}{4}\right) \end{pmatrix}\frac{1}{z} + \cdots\right].$$

In the previous series method it was proved rigorously that the resultant series are convergent. No such proof will be attempted in this section, and for a very good reason. These series need not converge. The following example will show such a case.

EXAMPLE 2. Solve

$$X' = \left[\begin{pmatrix} 0 & 1 \\ 0 & -1 \end{pmatrix} + \frac{1}{z}\begin{pmatrix} 0 & 0 \\ 0 & 1 \end{pmatrix} + \frac{1}{z^2}\begin{pmatrix} 0 & 0 \\ -1 & 0 \end{pmatrix}\right]X.$$

Solution. Letting

$$X = e^{\lambda z} z^{\mu} \sum_{k=0}^{\infty} C_k z^{-k},$$

we find that, from (6),

$$\begin{pmatrix} -\lambda & 1 \\ 0 & -\lambda - 1 \end{pmatrix} C_0 = 0.$$

Of the two roots $\lambda = 0$ and $\lambda = -1$, we select $\lambda = 0$ and find that

$$C_0 = a \begin{pmatrix} 1 \\ 0 \end{pmatrix}.$$

Then

$$\begin{pmatrix} 0 & 1 \\ 0 & -1 \end{pmatrix} C_1 = a \begin{pmatrix} \mu & 0 \\ 0 & \mu - 1 \end{pmatrix} \begin{pmatrix} 1 \\ 0 \end{pmatrix} = a \begin{pmatrix} \mu \\ 0 \end{pmatrix},$$

and for compatibility reasons, we find that $\mu = 0$ and

$$C_1 = a \begin{pmatrix} b \\ 0 \end{pmatrix}.$$

The general equation now is

$$\begin{pmatrix} 0 & 1 \\ 0 & -1 \end{pmatrix} C_k = (1 - k)\, C_{k-1} - A_1 C_{k-1} - A_2 C_{k-2},$$

and one can verify that we obtain

$$C_1 = a \begin{pmatrix} 1 \\ 0 \end{pmatrix}, \qquad C_k = a \begin{pmatrix} k! \\ -(k-1)\,(k-1)! \end{pmatrix}, \qquad k = 1, 2, \cdots.$$

The general solution now becomes

$$X = a \begin{pmatrix} 1 \\ 0 \end{pmatrix} + a \sum_{k=1}^{\infty} \begin{pmatrix} k! \\ -(k-1)\,(k-1)! \end{pmatrix} z^{-k},$$

and one sees that this series will converge for no finite value of z.

Series of the above type are called *asymptotic series*. The formal definition is now presented.

DEFINITION: The function $f(z)$ is said to have the asymptotic expansion

$$\sum_{k=0}^{\infty} a_k z^{-k}$$

as z approaches infinity if

$$\lim_{z\to\infty} z^n \left(f(z) - \sum_{k=0}^{n-1} a_k z^{-k}\right) = a_n, \qquad n = 0, 1, 2, \cdots.$$

For $n = 0$, the sum in the bracket must be interpreted as 0.

The solution (5) of (4) will be asymptotic in the sense that

$$\lim_{z\to\infty} z^n \left[e^{-\lambda z} z^{-\mu} X - \sum_{k=0}^{n-1} C_k z^{-k}\right] = C_n, \qquad n = 0, 1, \cdots.$$

More generally one can show that the equation

$$X' = z^r \sum_{k=0}^{\infty} A_k z^{-k} X$$

has an asymptotic solution of the form

$$X = e^{(a_0 z^{r+1} + a_1 z^r + \ldots + a_r z)} z^{\mu} \sum_{k=0}^{\infty} C_k z^{-k},$$

where the a_i are suitable scalars, provided A_0 has distinct eigenvalues. The steps are similar to the preceding cases and will not be shown here.

Example 3. Find an asymptotic solution for

$$y'' - zy = 0.$$

Solution. If one proceeds in a straightforward fashion and lets

$$x_1 = y,$$
$$x_2 = y',$$

one obtains

$$X' = \left[\begin{pmatrix} 0 & 0 \\ 1 & 0 \end{pmatrix} z + \begin{pmatrix} 0 & 1 \\ 0 & 0 \end{pmatrix}\right] X.$$

The leading matrix coefficient has a double eigenvalue at $\lambda = 0$, and the preceding methods fail. But a slight change in variables produces a system that can be solved. Let

$$x_1 = y,$$
$$x_2 = z^{-1/2} y',$$

so that

$$X' = \begin{pmatrix} 0 & z^{1/2} \\ z^{1/2} & -\dfrac{1}{2z} \end{pmatrix} X.$$

At this point it is convenient to introduce a new independent variable t by

$$z^{1/2} = t.$$

Then one obtains

$$\dot{X} = \left[t^2\begin{pmatrix}0 & 2\\ 2 & 0\end{pmatrix} + \frac{1}{t}\begin{pmatrix}0 & 0\\ 0 & -1\end{pmatrix}\right] X, \qquad \left(\dot{X} = \frac{d}{dt}X\right). \tag{8}$$

The leading matrix now has distinct eigenvalues, namely, $\lambda = \pm 2$.

We now let

$$X = e^{(a_0t^3+a_1t^2+a_2t)}\, t^\mu \sum_{k=0}^{\infty} C_k t^{-k}$$

and find from (8) that

$$e^{(a_0t^3+a_1t^2+a_2t)}\, t^\mu \sum_{k=0}^{\infty} \left(3a_0t^2 + 2a_1t + a_2 + \frac{\mu}{t} - \frac{k}{t}\right) C_k t^{-k}$$
$$= e^{(a_0t^3+a_1t^2+a_2t)}\, t^\mu \left[t^2\begin{pmatrix}0 & 2\\ 2 & 0\end{pmatrix} + \frac{1}{t}\begin{pmatrix}0 & 0\\ 0 & -1\end{pmatrix}\right] \sum_{k=0}^{\infty} C_k t^{-k}.$$

After dividing by $e^{(a_0t^3+a_1t^2+a_2t)}\, t^\mu$, one finds by equating the coefficients of t^2,

$$\begin{pmatrix}-3a_0 & 2\\ 2 & -3a_0\end{pmatrix} C_0 = 0,$$

so that $a_0 = \pm 2/3$. We shall choose $a_0 = -2/3$, and then

$$C_0 = a\begin{pmatrix}1\\ -1\end{pmatrix}.$$

Next we find, by equating coefficients of t, that

$$\begin{pmatrix}2 & 2\\ 2 & 2\end{pmatrix} C_1 = 2a_1C_0,$$

so that $a_1 = 0$. After making several such straightforward calculations, one finds

$$X = ae^{-2/3t^3}t^{-1/2}\left[\begin{pmatrix}1\\ -1\end{pmatrix} + \begin{pmatrix}-\frac{5}{48}\\ -\frac{7}{48}\end{pmatrix} t^{-3} + \cdots\right],$$

or in terms of the variable z,

$$X_1 = ae^{-2/3z^{3/2}}z^{-1/4}\left[\begin{pmatrix}1\\ -1\end{pmatrix} + \begin{pmatrix}-\frac{5}{48}\\ -\frac{7}{48}\end{pmatrix} z^{-3/2} + \cdots\right].$$

Similarly, if instead of $a_0 = -2/3$, the choice $a_0 = 2/3$ had been made, we should find

$$X_2 = be^{2/3z^{3/2}} z^{-1/4} \left[\begin{pmatrix} 1 \\ 1 \end{pmatrix} + \begin{pmatrix} \frac{5}{48} \\ -\frac{7}{48} \end{pmatrix} z^{-3/2} + \cdots \right].$$

These asymptotic properties are of great importance in physical problems. In many instances a condition on the solution will be that a solution tend to 0 as z tends to infinity. In the above example the most general solution will be

$$X = X_1 + X_2,$$

but coupled with the condition that X should vanish at infinity, the arbitrary constant $b = 0$, so that the only term retained is X_1. This still involves the arbitrary constant a, which must be determined from another condition.

As another application we return to the equation

$$X' = \frac{1}{z} \begin{pmatrix} 0 & 1 \\ -z^2 & 0 \end{pmatrix} X,$$

which is Bessel's equation with $n = 0$. In Section 3.3, Example 1, we found

$$X_1^o = a \sum_{k=0}^{\infty} \frac{(-1)^k}{4^k (k!)^2} \begin{pmatrix} 1 \\ 2k \end{pmatrix} z^{2k}.$$

In Example 2 we found a second solution:

$$X_2^o = b \begin{pmatrix} x_1 \int^z \frac{1}{tx_1^2}\, dt \\ x_2 \int^z \frac{t}{x_2^2}\, dt \end{pmatrix}.$$

The superscript ^{0} denotes the fact that these solutions are represented by expansion near the origin. In Section 3.4, Example 1, two solutions found were of the form

$$X_1^\infty = ce^{iz} z^{-1/2} \begin{pmatrix} i \\ -1 \end{pmatrix} + \cdots,$$

$$X_2^\infty = de^{-iz} z^{-1/2} \begin{pmatrix} -i \\ -1 \end{pmatrix} + \cdots,$$

where the superscript $^\infty$ denotes the fact that these are asymptotic solutions for large z. These four solutions are, of course, not linearly independent, since they come from a second-order system. Therefore there must exist relations of the form

$$X_1^0 = \alpha X_1^\infty + \beta X_2^\infty,$$

$$X_2^0 = \gamma X_1^\infty + \delta X_2^\infty,$$

but the determination of the constants α, β, γ, δ is very difficult and will not be attempted here. We can form two other asymptotic solutions:

$$X_3^\infty = e^{iz} z^{-1/2} \begin{pmatrix} i \\ -1 \end{pmatrix} + e^{-iz} z^{-1/2} \begin{pmatrix} -i \\ -1 \end{pmatrix} + \cdots$$

$$= z^{-1/2} \begin{pmatrix} -2 \sin z \\ -2 \cos z \end{pmatrix} + \cdots,$$

$$X_4^\infty = e^{iz} z^{-1/2} \begin{pmatrix} i \\ -1 \end{pmatrix} - e^{-iz} z^{-1/2} \begin{pmatrix} -i \\ -1 \end{pmatrix} + \cdots$$

$$= z^{-1/2} \begin{pmatrix} 2i \cos z \\ -2i \sin z \end{pmatrix} + \cdots,$$

by taking different combinations of X_1^∞ and X_2^∞. From these one can infer immediately that the components must have, for real values of z, an infinite number of real zeros. This fact will be of importance later in the discussion of boundary value problems.

Another class of asymptotic solutions arises in the solution of equations containing a parameter. Typical of this case is an equation like

$$X' = A(z, \mu)\, X, \tag{9}$$

where A depends on both the independent variable z and the parameter μ. In many cases of interest, A has an asymptotic expansion in the parameter μ of the form

$$A(z, \mu) = \mu \sum_{k=0}^{\infty} A_k(z)\, \mu^{-k}, \tag{10}$$

where the matrix coefficients are functions of z.

We shall first examine the simplest possible case, namely, the case where the expansion for A contains only a single term $\mu A_0(z)$ and the matrix A_0 is diagonal; that is,

$$A_0(z) = \begin{pmatrix} a_1(z) & & & 0 \\ & a_z(z) & & \\ & & \ddots & \\ 0 & & & a_n(z) \end{pmatrix}.$$

Then the system

$$X' = \mu A_0(z) X$$

can be rewritten as a system of n first-order equations:

$$x_i' = \mu a_i(z) x_i, \qquad i = 1, 2, \cdots n.$$

These evidently have the solutions

$$x_i = c_i \exp \left(\mu \int a_i(z)\, dz\right)$$

where the c_i are constants of integration.

We now return to the general equation (9). $A_0(z)$ will not be necessarily in diagonal form initially, but we shall assume that $A_0(z)$ has distinct eigenvalues; these of course, will be functions of z. In this case we shall be able to find a nonsingular matrix $T(z)$ such that $T^{-1}(z)\, A_0(z)\, T(z)$ will be diagonal. Then, by introducing a new dependent variable vector Y by

$$X = TY,$$

we have

$$T'Y + TY' = \mu \sum_{k=0}^{\infty} A_k \mu^{-k} TY$$

so that

$$Y' = \left[\mu T^{-1} A_0 T + (T^{-1} A_1 T - T^{-1} T') + \mu \sum_{k=2}^{\infty} T^{-1} A_k T \mu^{-k}\right] Y. \tag{11}$$

Now (11) is of type (9), where the expansion of A has the from (10), but the leading term is in diagonal form.

We shall now consider the equation

$$X' = \mu \sum_{k=0}^{\infty} A_k(z)\, \mu^{-k} X \tag{12}$$

and assume that $A_0(z)$ is diagonal and has distinct diagonal elements. In analogy to the previous series solutions, we shall now seek a solution in the form

$$X = \exp\left(\mu \int a_1(z)\, dz\right) \sum_{k=0}^{\infty} V_k(z)\, \mu^{-k}, \tag{13}$$

where the $V_k(z)$ are unknown vector coefficients. Inserting this in (12), we obtain

$$\mu a_1 \exp\left(\mu \int a_1\, dz\right) \sum_{k=0}^{\infty} V_k(z)\, \mu^{-k} + \exp\left(\mu \int a_1\, dz\right) \sum_{k=0}^{\infty} V_k'(z)\, \mu^{-k}$$

$$= \mu \exp\left(\mu \int a_1\, dz\right) \sum_{k=0}^{\infty} \mu^{-k} \sum_{l=0}^{k} A_{k-l} V_l(z).$$

By a comparison of corresponding coefficients of μ^{-k}, one now obtains the following recurrence formulas for the $V_k(z)$:

$$(A_0 - a_1 I)\, V_0 = 0,$$
$$(A_0 - a_1 I)\, V_k = V'_{k-1} - \sum_{l=0}^{k-1} A_{k-l} V_l, \qquad k = 1, 2, \cdots. \tag{14}$$

The first of these equations has a nonvanishing solution, since its determinant vanishes. The second equation also has a vanishing determinant, but V_0 has not been uniquely determined as yet and now is so chosen as to make the second equation compatible. Then V_1 will be determined, but not uniquely, until a compatibility condition is determined from the third equation, etc. Here, too, the series need not converge, but they will be asymptotic. n similar series solutions can be obtained by using all other eigenvalues of $A_0(z)$.

EXAMPLE 4. Obtain an asymptotic solution of

$$y'' + [\lambda - q(z)]\, y = 0$$

for large values of λ.

Solution. Let

$$\lambda = \mu^2,$$
$$y = x_1,$$
$$y' = \mu x_2,$$

so that we find

$$X' = \left[\mu \begin{pmatrix} 0 & 1 \\ -1 & 0 \end{pmatrix} + \frac{1}{\mu} \begin{pmatrix} 0 & 0 \\ q(z) & 0 \end{pmatrix}\right] X.$$

To diagonalize the leading matrix, we let

$$X = \begin{pmatrix} \dfrac{1}{2} & -\dfrac{i}{2} \\ -\dfrac{i}{2} & \dfrac{1}{2} \end{pmatrix} Y,$$

so that the equation becomes

$$Y' = \left[\mu A_0 + \frac{1}{\mu} A_2\right] Y,$$

where

$$A_0 = \begin{pmatrix} -i & 0 \\ 0 & i \end{pmatrix},$$
$$A_2 = \begin{pmatrix} iq/2 & q/2 \\ q/2 & -iq/2 \end{pmatrix}.$$

We now seek a solution in the form

$$Y = e^{-i\mu z} \sum_{k=0}^{\infty} V_k(z)\, \mu^{-k}$$

and find, as in (14),

$$(A_0 + iI)\, V_0 = 0,$$
$$(A_0 + iI)\, V_1 = V_0',$$
$$(A_0 + iI)\, V_k = V_{k-1}' - A_2 V_{k-2}, \qquad k = 2, 3, \cdots.$$

From the first of these we see that

$$\begin{pmatrix} 0 & 0 \\ 0 & 2i \end{pmatrix} V_0 = 0,$$

so that

$$V_0 = \begin{pmatrix} p_0(z) \\ 0 \end{pmatrix},$$

where $p_0(z)$ is an arbitrary function of z. The second equation becomes

$$\begin{pmatrix} 0 & 0 \\ 0 & 2i \end{pmatrix} V_1 = \begin{pmatrix} p_0'(z) \\ 0 \end{pmatrix},$$

from which one concludes that

$$p_0'(z) = 0$$

to make the equation compatible. Therefore $p_0(z)$ is a constant (say, a), and it follows that

$$V_0 = \begin{pmatrix} a \\ 0 \end{pmatrix}, \qquad V_1 = \begin{pmatrix} p_1(z) \\ 0 \end{pmatrix},$$

where $p_1(z)$ is an arbitrary function of z. Then the third equation becomes

$$\begin{pmatrix} 0 & 0 \\ 0 & 2i \end{pmatrix} V_2 = V_1' - \begin{pmatrix} iq/2 & q/2 \\ q/2 & -iq/2 \end{pmatrix} V_0$$

$$= \begin{pmatrix} p_1'(z) - iaq/2 \\ -aq/2 \end{pmatrix}.$$

The compatibility condition now leads to

$$p_1(z) = \frac{ia}{2} \int q\, dz,$$

so that

$$V_1 = \begin{pmatrix} \dfrac{ia}{2}\displaystyle\int q\,dz \\ 0 \end{pmatrix}, \qquad V_2 = \begin{pmatrix} p_2(z) \\ \dfrac{iaq}{4} \end{pmatrix},$$

where $p_2(z)$ is arbitrary. From the compatibility condition on the third equation one shows that

$$p_2'(z) = -\frac{a}{4} q \int q\,dz,$$

so that $p_2(z) = -(a/8)\left(\int q\,dz\right)^2$ and

$$Y = ae^{-i\mu z}\left[\begin{pmatrix} 1 \\ 0 \end{pmatrix} + \begin{pmatrix} \dfrac{i}{2}\displaystyle\int q\,dz \\ 0 \end{pmatrix}\mu^{-1} + \begin{pmatrix} -\dfrac{1}{8}\left(\displaystyle\int q\,dz\right)^2 \\ \dfrac{iq}{4} \end{pmatrix}\mu^{-2} + \cdots\right].$$

Finally, returning to the original differential equation, one finds that

$$y = ae^{-i\sqrt{\lambda}z}\left[\frac{1}{2} + \frac{i}{4}\int q\,dz\lambda^{-1/2} + \left(-\frac{1}{16}\left(\int q\,dz\right)^2 + \frac{q}{8}\right)\lambda^{-1} + \cdots\right].$$

One can generalize these methods to equations of the type

$$X' = \mu^r\left[\sum_{k=0}^{\infty} A_k(z)\,\mu^{-k}\right]X, \tag{15}$$

where r is an integer. In general, asymptotic solutions of type (13) for the case $r = 1$ exist only if the matrices $A_k(z)$ are analytic. More general cases become much more complicated and will not be discussed here.

Problems

1. Write the following equations as systems and solve for the first four non-vanishing terms in the series solution.

(a) $y'' + zy' = 0, \quad y(0) = 0, \; y'(0) = 1.$

(b) $y'' + (z-1)y' + y = 0, \quad y(0) = 1, \; y'(0) = 0.$

2. Reduce the following equations to the standard Euler form and obtain the general solution:

(a) $z^2y'' + zy' - 4y = 0.$

(b) $z^2y'' - 2y = 0.$

(c) $z^2y'' - 3zy' + 3y = 0.$

(d) $z^2y'' - zy' + y = 0.$

(e) $z^2y'' - 3zy' + 4y = 0.$

3. Show that if $X(z)$ represents a solution of the system

$$\frac{dX}{dz} = \frac{1}{z}\begin{pmatrix} 0 & 1 \\ n^2 - k^2z^2 & 0 \end{pmatrix} X,$$

then $Y(\zeta) = \zeta^{\rho\beta} X(\gamma\zeta^{\beta})$ satisfies

$$\frac{dY}{d\zeta} = \frac{1}{\zeta}\begin{pmatrix} \beta\rho & \beta \\ \beta n^2 - \beta k^2\gamma^2\zeta^{2\beta} & \beta\rho \end{pmatrix} Y.$$

4. Express the solution of

$$y'' + zy = 0$$

in terms of Bessel functions.

5. The equation

$$X' = A(z)\,X$$

is said to have a regular singular point at $z = \infty$, if after letting $z = 1/t$, the above reduces to the form

$$\frac{dX}{dt} = \frac{1}{t}\,B(t)\,X,$$

where $B(t)$ is analytic for small t. Show that

$$X' = \frac{1}{z}\,AX,$$

where A is constant, has a regular singular point at ∞.

6. Show that the only equation with exactly two regular singular points of which one is at infinity is given by

$$X' = \frac{1}{z}\,AX,$$

where A is constant. Show that it would be impossible for the equation to have only two regular singular points, neither of which is at infinity.

7. Show that no equation can have only one regular singular point.

8. Show that the only equation with only three regular singular points at the three points $z = 0$, $z = 1$, $z = \infty$ is the equation

$$X' = \left(\frac{A_1}{z} + \frac{A_2}{z-1}\right) X,$$

where A_1 and A_2 are constant.

9. Supply a proof of (8) in Section 3.3.

10. Find a series solution for

$$zy'' - y' + 4z^3 y = 0.$$

Obtain a closed form solution, by summing the resultant series.

11. Show that

$$z(1-z)\,y'' + [c - (a+b+1)\,z]\,y' - aby = 0$$

has three regular singular points at $z = 0$, $z = 1$, $z = \infty$. Verify that the following represents a series solution

$$F(a, b; c; z) = \sum_{n=0}^{\infty} \frac{a_{(n)} b_{(n)}}{c_{(n)} n!} z^n,$$

where $a_{(n)} = a(a+1)(a+2)\cdots(a+n-1)$. This solution is known as a hypergeometric function. The differential equation is known as the hypergeometric differential equation.

12. Prove the following, either by use of the series or from the differential equation:

$$F(-n, 1; 1; -z) = (1+z)^n,$$

$$zF(1, 1; 2; -z) = \ln(1+z),$$

$$zF\left(\frac{1}{2}, \frac{1}{2}; \frac{3}{2}; z^2\right) = \sin^{-1} z,$$

$$F(a, b; 2b; z) = \left(1 - \frac{z}{2}\right)^{-a} F\left(\frac{a}{2}, \frac{a+1}{2}; b + \frac{1}{2}; \left(\frac{z}{2-z}\right)^2\right).$$

13. If z is replaced by z/b and b is allowed to become infinite, the hypergeometric differential equation degenerates into the confluent hypergeometric equation

$$zy'' + (c - z)\,y' - ay = 0.$$

This equation has a regular singular point at $z = 0$, and an irregular singular point at $z = \infty$. The hypergeometric series reduces to

$${}_1F_1(a; c; z) = \sum_{0}^{\infty} \frac{a_{(n)}}{c_{(n)} n!} z^n.$$

Show that

$$_1F_1(a; a; z) = e^z,$$

$$_1F_1\left(n + \frac{1}{2}, 2n + 1, 2iz\right) = \frac{e^{iz} n!\, J_n(z)}{\left(\frac{z}{2}\right)^n},$$

where $J_n(z)$ is a Bessel function. It satisfies the differential equation

$$y'' + \frac{1}{z} y' + \left(1 - \frac{n^2}{z^2}\right) y = 0.$$

14. Find asymptotic series solutions for large z for

$$z^2 y'' + (1 - z^2)\, y' - (2 + z)\, y = 0.$$

15. Find asymptotic series solutions for large z for the confluent hypergeometric equation

$$zy'' + (c - z)\, y' - ay = 0.$$

16. Find asymptotic series solutions for large z for

$$X' = \left[\begin{pmatrix} 0 & 1 \\ -4 & 0 \end{pmatrix} + \frac{1}{z^2}\begin{pmatrix} 0 & 0 \\ n^2 & 0 \end{pmatrix}\right] X.$$

17. In Example 1 of Section 3.4 it was shown that the asymptotic series solution is a finite series for $n^2 = 1/4$. Show that this is always the case when $n^2 = m^2/4$, where m is an odd integer.

18. Show that the equation

$$X' = z^r \sum_0^\infty A_k z^{-k} X,$$

where A_0 has distinct eigenvalues, has asymptotic solutions of the form

$$X = e^{(a_0 z^{r+1} + a_1 z^r + \cdots + a_r z)}\, z^u \sum_{k=0}^\infty C_k z^{-k},$$

where $a_0, a_1, \cdots, a_r, \mu$ are scalars, and C_k constant vectors.

19. Obtain asymptotic series solutions for large z, for

(a) $zy'' - (1 + z)\, y' + zy = 0.$

(b) $zy'' - (1 + z)\, y' + z^3 y = 0.$

20. Find an asymptotic series solution for large μ for

$$y'' - \mu e^z y' - \mu y = 0.$$

4 Boundary Value Problems

4.1 INTRODUCTION

In many physical problems one encounters second-order differential equations of the type

$$p_0(z)\,y'' + p_1(z)\,y' + (p_2(z) + \lambda p_3(z))\,y = 0 \tag{1}$$

with the auxiliary homogeneous boundary conditions

$$\begin{aligned} a_0 y(z_0) + a_1 y'(z_0) + a_2 y(z_1) + a_3 y'(z_1) &= 0, \\ b_0 y(z_0) + b_1 y'(z_0) + b_2 y(z_1) + b_3 y'(z_1) &= 0, \qquad z_0 < z_1. \end{aligned} \tag{2}$$

The solution is to be determined in the interval (z_0, z_1). The boundary conditions (2) are such that in general no nonvanishing solution of (1) can satisfy both boundary conditions (2), unless the parameter λ in (1) takes on special values.

Some typical physical problems that lead to equations of this type will now be discussed. In dealing with vibrations of circular membranes, one has to solve the partial differential equations

$$\frac{\partial^2}{\partial r^2} u(r, t) + \frac{1}{r}\frac{\partial}{\partial r} u(r, t) = c^2 \frac{\partial^2}{\partial t^2} u(r, t) \tag{3}$$

with the boundary condition

$$u(a, t) = 0 \tag{4}$$

and the initial conditions

$$\begin{aligned} u(r, 0) &= f(r), \\ \frac{\partial}{\partial t} u(r, 0) &= g(r). \end{aligned} \tag{5}$$

Here r is the radial coordinate of the membrane and t is a time dimension. The radius of the membrane is taken to be a, and (4) demands that the outer ring of the membrane be constrained to lie in the plane $u = 0$, the rest position; (5) prescribes the initial displacement and velocity of the membrane.

The standard approach to such problems is by the method of separation of variables. That is, the solution of (3) is assumed to be a product of two functions; one a function of r only, and the other is a function of t only. Then

$$u(r, t) = R(r)\,T(t). \tag{6}$$

If (6) is inserted in (3) and then the expression is divided by RT, one obtains

$$\frac{\dfrac{d^2R}{dr^2} + \dfrac{1}{r}\dfrac{dR}{dr}}{R} = c^2 \frac{\dfrac{d^2T}{dt^2}}{T}.$$

Since one side of the preceding equation is a function of r only and the other of t only, and these are independent variables, one is led to the conclusion that both are equal to the same constant, say, $-\lambda$. Thus one obtains the two ordinary differential equations:

$$\begin{aligned} \frac{d^2T}{dt^2} + \frac{\lambda}{c^2} T &= 0, \\ \frac{d^2R}{dr^2} + \frac{1}{r}\frac{dR}{dr} + \lambda R &= 0. \end{aligned} \tag{7}$$

On the second equation of (7) one has to impose the condition

$$R(a) = 0$$

from (4), and from the symmetry of the membrane one expects the slope R' to vanish at the center; that is,

$$R'(0) = 0.$$

This is now an equation of type (1) with boundary conditions of type (2) at $r = 0$ and $r = a$.

Another physical problem involves heat flow in a thin circular ring. This situation is governed by the partial differential equation

$$\frac{\partial^2}{\partial \theta^2} u(\theta, t) = \alpha \frac{\partial}{\partial t} u(\theta, t) \tag{8}$$

where t is again a time coordinate and θ describes the annular ring. Proceeding by the method of separation of variables we let

$$u(\theta, t) = y(\theta)\, T(t)$$

and find that

$$\frac{\dfrac{d^2}{d\theta^2} y}{y} = \alpha \frac{\dfrac{d}{dt} T}{T}$$

As before, each of the above functions must equal the same constant (say, k) and we find that

$$\begin{aligned} \frac{d}{dt} T &= \frac{k}{\alpha} T \\ \frac{d^2}{d\theta^2} y - ky &= 0. \end{aligned}$$

One of the auxiliary conditions here is the initial condition

$$u(\theta, 0) = f(\theta).$$

There are no boundaries in this case, but one has to impose the periodicity condition that

$$u(0, t) - u(2\pi, t) = 0,$$

$$\frac{\partial}{\partial\theta} u(\theta, t)\Big|_{\theta=0} - \frac{\partial}{\partial\theta} u(\theta, t)\Big|_{\theta=2\pi} = 0.$$

This leads to the boundary conditions on y:

$$y(0) - y(2\pi) = 0,$$
$$y'(0) - y'(2\pi) = 0.$$

Here again one finds a differential equation of type (1) with boundary conditions (2).

4.2 NORMAL FORMS

We now return to (1) of Section 4.1 and assume that the coefficients $p_0(z)$ and $p_3(z)$ are positive and that $p_0(z)$, $p_1(z)$, and $p_3(z)$ are twice differentiable. If we now let

$$p(z) = \exp \int \frac{p_1(z)}{p_0(z)}\, dz,$$

$$r(z) = \frac{p_2(z)p(z)}{p_0(z)},$$

$$g(z) = \frac{p_3(z)p(z)}{p_0(z)},$$

and multiply (1) of Section 4.1 by $p(z)/p_0(z)$, we obtain

$$\frac{d}{dz} p(z) \frac{dy}{dz} + (r(z) + \lambda g(z))\, y = 0, \tag{1}$$

which, as will be seen later, is a more convenient form. This form will be referred to as the self-adjoint form in analogy to the case of self-adjoint transformations. The reasons for this analogy will emerge shortly.

Now (1) can be simplified still further if one introduces the new independent variable

$$\zeta = \int \frac{dz}{p(z)},$$

so that (1) becomes

$$\frac{d^2}{d\zeta^2} y + (p(z)\,r(z) + \lambda p(z)\,g(z))\,y = 0. \tag{2}$$

Still another reduction can be obtained by letting

$$y = k(\zeta)\,u(\xi), \qquad \xi = \int \frac{d\zeta}{k^2(\zeta)},$$

thus introducing both a new dependent and independent variable. Now

$$\frac{dy}{d\zeta} = u\,\frac{dk}{d\zeta} + k\,\frac{du}{d\xi}\,\frac{d\xi}{d\zeta} = u\,\frac{dk}{d\zeta} + \frac{1}{k}\,\frac{du}{d\xi},$$

$$\frac{d^2y}{d\zeta^2} = u\,\frac{d^2k}{d\zeta^2} + \frac{1}{k^3}\,\frac{d^2u}{d\xi^2},$$

so that (2) becomes

$$\frac{1}{k^3}\,\frac{d^2u}{d\xi^2} + \left(p(z)\,r(z) + \frac{1}{k}\,\frac{d^2k}{d\zeta^2} + \lambda p(z)\,g(z)\right) ku = 0.$$

If k is now chosen so that

$$k^4 p(z)\,g(z) = 1,$$

we obtain the still simpler normal form

$$u'' + (-q(\xi) + \lambda)\,u = 0, \tag{3}$$

where

$$u'' = \frac{d^2u}{d\xi^2}$$

and

$$-q(\xi) = k^4\left(p(z)\,r(z) + \frac{d^2k}{d\zeta^2}\right)$$

when the right side is expressed as a function of ξ.

EXAMPLE. Transform the equation

$$y'' + \frac{1}{z}y' + \left(\lambda - \frac{N^2}{z^2}\right)y = 0$$

into normal forms (3).

Solution. After multiplying by

$$\exp \int \frac{1}{z}\,dz = z,$$

the equation takes the form

$$\frac{d}{dz} z \frac{dy}{dz} + \left(\lambda z - \frac{N^2}{z}\right) y = 0$$

as in (1). Let

$$\zeta = \int \frac{dz}{z} = \ln z,$$

so that, as in (2),

$$\frac{d^2y}{d\zeta^2} + (\lambda e^{2\zeta} - N^2) y = 0.$$

Next

$$k^4 e^{2\zeta} = 1,$$

so that

$$y = e^{-\zeta/2} u(\xi), \qquad \xi = \int e^{\zeta}\, d\zeta = e^{\zeta},$$

and (3) becomes

$$u''(\xi) + \left(\lambda - \frac{N^2 - \frac{1}{4}}{\xi^2}\right) u(\xi) = 0.$$

For the special case where $N^2 = 1/4$, this reduces to an equation with constant coefficients.

Equation (3) is sometimes referred to as the Liouville normal form. If one starts with (1) and boundary conditions (2) in Section 4.1 and transforms (1) into the Liouville normal form (3), then the boundary conditions are transformed into new homogeneous boundary conditions of type (2) of Section 4.1.

4.3 EQUATION $u'' + \lambda u = 0$

The equation

$$u'' + \lambda u = 0 \tag{1}$$

can be solved by the methods of Chapter 2. Rather than solve the equation in terms of well-known exponential or trigonometric functions, we shall deduce the important properties of the solutions directly. The equation has analytic coefficients and it must therefore possess solutions expressible in terms of power series. Two series representing linearly independent solutions are

$$u_1 = \sum_{n=0}^{\infty} \frac{(\sqrt{-\lambda}\, z)^n}{n!},$$

$$u_2 = \sum_{n=0}^{\infty} \frac{(-\sqrt{-\lambda}\, z)^n}{n!}. \tag{2}$$

According to the theorems of Chapter 2 these series must converge in all domains where the analytic coefficients of (1) converge. Since in this case the coefficients are constant, so that their series converge for all finite values of z, it follows that the series (2) converge for all finite values of z. This can also be confirmed by direct tests like the ratio test.

We first consider the case where the parameter λ is negative, say,

$$\lambda = -\mu^2.$$

Then

$$u_1 = \sum_{n=0}^{\infty} \frac{(\mu z)^n}{n!},$$

and evidently $u_1(0) = u_2(0) = 1$.

One observes that

$$u_1 > 1 \qquad \text{for } \mu z > 0,$$

and since in this case u_1 is a sum of positive and increasing terms, it follows that u_1 is an increasing function of μz. By term-by-term differentiation, justified by the uniform convergence of these series in every finite interval, it follows that

$$u_1' = \sum_{n=0}^{\infty} \frac{n\mu^n z^{n-1}}{n!} = \mu \sum_{n=1}^{\infty} \frac{(\mu z)^{n-1}}{(n-1)!} = \mu \sum_{n=0}^{\infty} \frac{(\mu z)^n}{n!} = \mu u_1,$$

and similarly,

$$u_2' = -\mu u_2.$$

From the series it also follows that

$$u_2(-z) = u_1(z).$$

We now consider the function

$$v(z) = u_1(z + a) - u_1(z)\, u_1(a),$$

where a is a fixed constant. By differentiation we find

$$v'(z) = u_1'(z + a) - u_1'(z)\, u_1(a) = \mu u_1(z + a) - \mu u_1(z)\, u_1(a) = \mu v(z).$$

Therefore the function $v(z)$ satisfies the first-order differential equation

$$v'(z) = \mu v(z)$$

with the initial value

$$v(0) = u_1(a) - u_1(a) = 0.$$

An equation of this type must have a unique solution, and since a solution is $v = 0$, it follows that

$$v(z) = 0,$$

so that

$$u_1(z + a) = u_1(z)\, u_1(a).$$

Since this must hold for all finite values of z and a, we can let $a = -z$ so that

$$u_1(z)\, u_1(-z) = u_1(0) = 1.$$

We deduce from this that since $u_1(z)$ is an increasing function of z for positive values of z, it must be increasing for all real values of z. Also, in view of the fact that $u_1(z)$ takes on only finite values for finite values of z, it follows that $u_1(z)$ can never vanish, for if at $z = x$,

$$u_1(x) = 0,$$

it would follow that $u_1(-x)$ is no longer finite. Hence we have that

$$0 < u_1(z)$$

for all finite real z, and it is an increasing function. From this we conclude that the equation

$$u_1(z) = b,$$

where b is a real constant, can have only one real solution.

Similar results hold for $u_2(z) = u_1(-z)$ except that $u_2(z)$ is a decreasing function. If a knowledge of exponential function were presupposed, all these conclusions would be clear, since we have

$$u_1(z) = e^{\mu z},$$
$$u_2(-z) = e^{-\mu z},$$

but we have derived the most important properties of these functions directly from the differential equation.

We shall now return to the differential equation (1) and solutions (2), but shall consider the case where λ is positive, so that

$$\sqrt{-\lambda} = i\sqrt{\lambda}.$$

Since

$$u_1(z) = e^{i\sqrt{\lambda}z},$$

$$u_2(z) = e^{-i\sqrt{\lambda}z},$$

we can construct other solutions by taking linear combinations of these. Two solutions that will play an important role in the sequel are

$$\cos \sqrt{\lambda}\, z = \frac{1}{2}(e^{i\sqrt{\lambda}z} + e^{-i\sqrt{\lambda}z}),$$

$$\sin \sqrt{\lambda}\, z = \frac{1}{2i}(e^{i\sqrt{\lambda}z} - e^{-i\sqrt{\lambda}z}),$$

and from the series (2) it follows that

$$\cos \sqrt{\lambda}\, z = \sum_{n=0}^{\infty} \frac{(-1)^n (\sqrt{\lambda}\, z)^{2n}}{(2n)!}$$

$$\sin \sqrt{\lambda}\, z = \sqrt{\lambda}\, z \sum_{n=0}^{\infty} \frac{(-1)^n (\sqrt{\lambda}\, z)^{2n}}{(2n+1)!} \tag{3}$$

It should be observed that the names "cos" and "sin" are used here in anticipation of the fact that the series (3) do indeed represent the familiar trigonometric functions, but none of their properties will be presupposed. All properties needed will be derived from (1) and (3). First we observe that

$$\cos 0 = 1,$$
$$\sin 0 = 0,$$

and by differentiation,

$$\frac{d}{dz} \cos \sqrt{\lambda}\, z = \sum_{n=0}^{\infty} \frac{2n(-1)^n (\sqrt{\lambda})^{2n} z^{2n-1}}{(2n)!} = -\lambda z \sum_{n=0}^{\infty} \frac{(-1)^n (\sqrt{\lambda}\, z)^{2n}}{(2n+1)!}$$
$$= -\sqrt{\lambda} \sin \sqrt{\lambda}\, z,$$

and similarly,

$$\frac{d}{dz} \sin \sqrt{\lambda}\, z = \sqrt{\lambda} \cos \sqrt{\lambda}\, z.$$

From these we can deduce that these functions are characterized as being those solutions of (1) with the initial values

$$\cos 0 = 1, \qquad \sin 0 = 0,$$

$$\left.\frac{d}{dz} \cos \sqrt{\lambda}\, z \right|_{z=0} = 0,$$

$$\left.\frac{d}{dz} \sin \sqrt{\lambda} z \right|_{z=0} = \sqrt{\lambda}.$$

We shall now examine the function

$$v(z) = \cos \sqrt{\lambda}(z + a),$$

where a is a fixed quantity. Then

$$v'(z) = -\sqrt{\lambda} \sin \sqrt{\lambda}(z + a),$$

$$v''(z) = -\lambda \cos \sqrt{\lambda}(z + a).$$

Therefore $v(z)$ satisfies the differential equation (1):

$$v'' + \lambda v = 0$$

with initial values

$$v(0) = \cos \sqrt{\lambda}\, a,$$

$$v'(0) = -\sqrt{\lambda} \sin \sqrt{\lambda}\, a.$$

But from (3) one can construct a solution of (1) with these initial values, namely,

$$v(z) = \cos \sqrt{\lambda}\, a \cos \sqrt{\lambda}\, z - \sin \sqrt{\lambda}\, a \sin \sqrt{\lambda}\, z.$$

Since a second-order differential equation of type (1) with two initial conditions must have a unique solution, we conclude that

$$\cos \sqrt{\lambda}(z + a) = \cos \sqrt{\lambda}\, a \cos \sqrt{\lambda}\, z - \sin \sqrt{\lambda}\, a \sin \sqrt{\lambda}\, z,$$

and by differentiating the above and dividing by $-\sqrt{\lambda}$, we have

$$\sin \sqrt{\lambda}(z + a) = \cos \sqrt{\lambda}\, a \sin \sqrt{\lambda}\, z + \sin \sqrt{\lambda}\, a \cos \sqrt{\lambda}\, z.$$

These are known as the addition theorems for the cosine and sine functions, respectively.

From (3) one sees that the cosine is an even function and the sine is an odd function of z; that is,

$$\cos (-\sqrt{\lambda}\, z) = \cos \sqrt{\lambda}\, z,$$

$$\sin (-\sqrt{\lambda}\, z) = -\sin \sqrt{\lambda}\, z.$$

If we now let $a = -z$ in the addition theorem for the cosine function, we have

$$1 = \cos^2 \sqrt{\lambda}\, z + \sin^2 \sqrt{\lambda}\, z,$$

which is the Pythagorean theorem. From it one concludes that for real z,

$$-1 \leqslant \cos \sqrt{\lambda}\, z \leqslant 1,$$

$$-1 \leqslant \sin \sqrt{\lambda}\, z \leqslant 1.$$

We shall now show that there is a point $z_1 > 0$ such that

$$\cos \sqrt{\lambda}\, z_1 = 0.$$

To do so, we observe that

$$\cos \sqrt{\lambda}\, z \leqslant 1.$$

By an integration it follows that

$$\int_0^z \cos \sqrt{\lambda}\, t \, dt \leqslant z, \qquad z \geqslant 0,$$

so that

$$\sin \sqrt{\lambda}\, z \leqslant \sqrt{\lambda}\, z.$$

By repeated integrations it follows that

$$\cos \sqrt{\lambda}\, z \geqslant 1 - \frac{\lambda z^2}{2},$$

$$\sin \sqrt{\lambda}\, z \geqslant \sqrt{\lambda}\, z - \frac{(\sqrt{\lambda}\, z)^3}{6},$$

$$\cos \sqrt{\lambda}\, z \leqslant 1 - \frac{\lambda z^2}{2} + \frac{\lambda^2 z^4}{24}.$$

From the last inequality it follows that for some positive value of $\sqrt{\lambda}\, z$ (say, $\sqrt{\lambda} z_1$),

$$\cos \sqrt{\lambda}\, z_1 = 0.$$

From the Pythagorean theorem we see that at the point where

$$\cos \sqrt{\lambda}\, z_1 = 0,$$

it follows that

$$\sin^2 \sqrt{\lambda}\, z_1 = 1.$$

The right side of the last inequality is a decreasing function. Therefore its slope is negative, so that

$$-\sqrt{\lambda}\, z_1 + \frac{(\sqrt{\lambda}\, z_1)^3}{6} < 0.$$

Combining this result with the third inequality, we see that $\sin\sqrt{\lambda}\,z_1$ must be positive, so that

$$\sin\sqrt{\lambda}\,z_1 = 1.$$

By applying the addition theorems we now find

$$\cos 2\sqrt{\lambda}\,z_1 = \cos^2\sqrt{\lambda}\,z_1 - \sin^2\sqrt{\lambda}\,z_1 = -1,$$
$$\sin 2\sqrt{\lambda}\,z_1 = 2\sin\sqrt{\lambda}\,z_1\cos\sqrt{\lambda}\,z_1 = 0,$$

and

$$\cos 4\sqrt{\lambda}\,z_1 = \cos^2 2\sqrt{\lambda}z_1 - \sin^2 2\sqrt{\lambda}\,z_1 = 1,$$
$$\sin 4\sqrt{\lambda}\,z_1 = 2\sin 2\sqrt{\lambda}\,z_1\cos 2\sqrt{\lambda}\,z_1 = 0.$$

It follows that

$$\cos(\sqrt{\lambda}\,z + 4\sqrt{\lambda}\,z_1) = \cos\sqrt{\lambda}\,z\cos 4\sqrt{\lambda}\,z_1 - \sin\sqrt{\lambda}\,z\sin 4\sqrt{\lambda}\,z_1 = \cos\sqrt{\lambda}z,$$

$$\sin(\sqrt{\lambda}\,z + 4\sqrt{\lambda}\,z_1) = \sin\sqrt{\lambda}\,z\cos 4\sqrt{\lambda}\,z_1 + \cos\sqrt{\lambda}\,z\sin 4\sqrt{\lambda}\,z_1 = \sin\sqrt{\lambda}\,z.$$

The conclusion of these results is that if $u(z)$ is any solution of

$$u'' + \lambda u = 0,$$

then

$$u(z + 4z_1) = u(z).$$

In other words, u is a periodic function of period $4z_1$.

The number $\sqrt{\lambda}\,z_1$ can be given numerous other interpretations. We now define the function

$$\tan\sqrt{\lambda}\,z = \frac{\sin\sqrt{\lambda}\,z}{\cos\sqrt{\lambda}\,z}$$

and see that

$$\frac{d}{dz}\tan\sqrt{\lambda}\,z = \sqrt{\lambda}\,\frac{\cos^2\sqrt{\lambda}\,z + \sin^2\sqrt{\lambda}\,z}{\cos^2\sqrt{\lambda}\,z} = \sqrt{\lambda}(1 + \tan^2\sqrt{\lambda}\,z).$$

Then, if we let

$$\tan\sqrt{\lambda}\,z = y,$$

we have

$$\frac{dy}{1 + y^2} = \sqrt{\lambda}\,dz,$$

and since

$$\tan 0 = 0,$$

$$\lim_{z \to z_1} \tan \sqrt{\lambda}\, z = \infty,$$

we obtain by integration:

$$\sqrt{\lambda}\, z_1 = \int_0^\infty \frac{dy}{1 + y^2}.$$

The integral on the right is a convergent definite integral and therefore represents the finite number $\sqrt{\lambda}\, z_1$.

From this integral we can obtain another interpretation of $\sqrt{\lambda}\, z_1$. We suppose that the function $V(t)$ is a convex differentiable function of t, and we suppose that there are two values t_0 and t_1 such that, at t_0, V' vanishes; at t_1, V' is infinite. Then we let

$$y = V'(t),$$

so that

$$\frac{dy}{1 + y^2} = \frac{V''\, dt}{1 + V'^2} = \frac{V''\, ds}{(1 + V'^2)^{3/2}} = \frac{ds}{\rho},$$

where ds is the arc length along the curve and ρ is its radius of curvature. Then

$$\sqrt{\lambda}\, z_1 = \int_0^\infty \frac{dy}{1 + y^2} = \int_{t_0}^{t_1} \frac{ds}{\rho}.$$

But the integral on the right represents the angle between the tangents at t_0 and t_1, which is known to be $\pi/2$. Hence

$$\sqrt{\lambda}\, z_1 = \frac{\pi}{2}$$

and the solutions of (1) have period $2\pi/\sqrt{\lambda}$.

From the preceding results one can draw conclusions regarding the solvability of equations of the type

$$a \cos \sqrt{\lambda}\, z + b \sin \sqrt{\lambda}\, z = c. \tag{4}$$

First we consider the simultaneous equations

$$a = r \cos \theta,$$
$$b = r \sin \theta,$$

to be solved for r and θ. By use of the Pythagorean theorem, θ can be eliminated, and one finds that

$$r = \sqrt{a^2 + b^2}.$$

Now $\cos \theta$ takes on all values between -1 and 1, so that we can find a value of θ such that

$$\cos \theta = \frac{a}{r}.$$

But since $\cos \theta$ is an even function, we also have

$$\cos(-\theta) = \frac{a}{r}.$$

From

$$\sin^2 \theta = 1 - \cos^2 \theta = 1 - \frac{a^2}{r^2} = \frac{b^2}{r^2},$$

we have either

$$\sin \theta = \frac{b}{r} \quad \text{or} \quad \sin(-\theta) = -\sin \theta = \frac{b}{r}.$$

Hence either θ or $-\theta$ is a solution. By replacing a and b in (4) in terms of r and θ, one obtains

$$\cos \theta \cos \sqrt{\lambda}\, z + \sin \theta \sin \sqrt{\lambda}\, z = \frac{c}{r}.$$

From the addition theorem this reduces to

$$\cos(\sqrt{\lambda}\, z - \theta) = \frac{c}{r}.$$

It follows that (4) can have, dealing only in the real domain, solutions if and only if

$$\left| \frac{c}{r} \right| \leqslant 1.$$

Then, if one such solution is given by

$$\sqrt{\lambda}\, z = \theta + \zeta,$$

one obtains from the periodicity of the cosine function:

$$\sqrt{\lambda}\, z = \theta + \zeta + 2n\pi, \quad n \text{ integral}.$$

From the evenness of the cosine it follows that a second family of solutions is given by

$$\sqrt{\lambda}\, z = \theta - \zeta + 2n\pi, \quad n \text{ integral},$$

and these two sets of solutions furnish all solutions of (4).

We now turn our attention to the boundary value problem

$$y'' + \lambda y = 0, \tag{5}$$

$$\begin{aligned} a_0 y(0) + a_1 y'(0) + a_2 y(\pi) + a_3 y'(\pi) &= 0, \\ b_0 y(0) + b_1 y'(0) + b_2 y(\pi) + b_3 y'(\pi) &= 0, \end{aligned} \tag{6}$$

where all a_i and b_i are real. We wish to discover whether or not, and under what conditions, we can find solutions of (5) satisfying (6) as well. The most general solution of (5) can now be expressed in the form

$$y = c_1 \sin \sqrt{\lambda}\, z + c_2 \cos \sqrt{\lambda}\, z, \tag{7}$$

where c_1 and c_2 are arbitrary constants. From (6) we obtain the simultaneous equations for c_1 and c_2:

$$\begin{aligned} &(a_1 \sqrt{\lambda} + a_2 \sin \sqrt{\lambda}\pi + a_3 \sqrt{\lambda} \cos \sqrt{\lambda}\pi)\, c_1 \\ &\qquad\qquad + (a_0 + a_2 \cos \sqrt{\lambda}\pi - a_3 \sqrt{\lambda} \sin \sqrt{\lambda}\pi)\, c_2 = 0, \\ &(b_1 \sqrt{\lambda} + b_2 \sin \sqrt{\lambda}\pi + b_3 \sqrt{\lambda} \cos \sqrt{\lambda}\pi)\, c_1 \\ &\qquad\qquad + (b_0 + b_2 \cos \sqrt{\lambda}\pi - b_3 \sqrt{\lambda} \sin \sqrt{\lambda}\pi)\, c_2 = 0. \end{aligned} \tag{8}$$

In order for this homogeneous system to have nonvanishing solutions for c_1 and c_2, it is necessary and sufficient that the determinant of the system vanish. Let

$$= \begin{vmatrix} a_1 \sqrt{\lambda} + a_2 \sin \sqrt{\lambda}\pi + a_3 \sqrt{\lambda} \cos \sqrt{\lambda}\pi & a_0 + a_2 \cos \sqrt{\lambda}\pi - a_3 \sqrt{\lambda} \sin \sqrt{\lambda}\pi \\ b_1 \sqrt{\lambda} + b_2 \sin \sqrt{\lambda}\,\pi + b_3 \sqrt{\lambda} \cos \sqrt{\lambda}\,\pi & b_0 + b_2 \cos \sqrt{\lambda}\pi - b_3 \sqrt{\lambda} \sin \sqrt{\lambda}\pi \end{vmatrix} \tag{9}$$

The dependence of this so-called characteristic determinant on λ is brought out by writing $\Delta(\lambda)$. In general this determinant can vanish only for special values of λ. But if, for example,

$$a_i = kb_i, \qquad i = 0, 1, 2, 3,$$

then $\Delta(\lambda)$ vanishes identically. In this case one can determine c_1 and c_2 from (8) so as to obtain nonvanishing solutions of (5) satisfying (6) for all λ. We shall disregard such cases and concern ourselves exclusively with those where $\Delta(\lambda)$ does not vanish identically.

In general the solutions of

$$\Delta(\lambda) = 0 \tag{10}$$

will be complex numbers. But restricting ourselves to boundary conditions less general than (6), we can show that all roots of (10) are real. Such cases will be referred to as self-adjoint problems, for reasons that will be given shortly. These are, in the first place, the most important in physical problems and, secondly, are the only cases for which a general mathematical theory can be constructed. There are important nonselfadjoint problems, but these must be treated as special cases. There are two important self-adjoint cases, one in which the boundary conditions are separated, that is,

$$\begin{aligned} a_0 y(0) + a_1 y'(0) &= 0, \\ b_0 y(\pi) + b_1 y'(\pi) &= 0, \end{aligned} \tag{11}$$

and periodicity boundary conditions

$$\begin{aligned} y(0) - y(\pi) &= 0, \\ y'(0) - y'(\pi) &= 0. \end{aligned} \tag{12}$$

Corresponding to (11) we find that

$$\begin{aligned} \Delta(\lambda) &= \begin{vmatrix} a_1 \sqrt{\lambda} & a_0 \\ b_0 \sin \sqrt{\lambda}\,\pi + b_1 \sqrt{\lambda} \cos \sqrt{\lambda}\,\pi & b_0 \cos \sqrt{\lambda}\,\pi - b_1 \sqrt{\lambda} \sin \sqrt{\lambda}\pi \end{vmatrix} \\ &= (a_1 b_0 - a_0 b_1) \sqrt{\lambda} \cos \sqrt{\lambda}\pi - (a_0 b_0 + a_1 b_1 \lambda) \sin \sqrt{\lambda}\,\pi, \end{aligned} \tag{13}$$

and corresponding to (12),

$$\begin{aligned} \Delta(\lambda) &= \begin{vmatrix} -\sin \sqrt{\lambda}\,\pi & 1 - \cos \sqrt{\lambda}\pi \\ \sqrt{\lambda} - \sqrt{\lambda} \cos \sqrt{\lambda}\,\pi & \sqrt{\lambda} \sin \sqrt{\lambda}\,\pi \end{vmatrix} \\ &= -2\sqrt{\lambda}\,(1 - \cos \sqrt{\lambda}\pi) = -4\sqrt{\lambda} \sin^2 \tfrac{1}{2} \sqrt{\lambda}\pi. \end{aligned} \tag{14}$$

One can show that all zeros of $\Delta(\lambda)$ as given by (13) and (14) must be real. To do so, one assumes that λ is a complex zero. It follows that $\bar{\lambda}$ is also a zero, since all coefficients in (13) and (14) are real. Then, if y satisfies

$$y'' + \lambda y = 0$$

with the boundary conditions (11) or (12), it follows that $\bar{y}$ satisfies

$$\bar{y}'' + \bar{\lambda}\bar{y} = 0$$

with the same boundary conditions. Then

$$\bar{y}(y'' + \lambda y) - y(\bar{y}'' + \bar{\lambda}\bar{y}) = 0,$$

and by rearranging the terms and integrating over the interval $(0, \pi)$, one finds

$$\int_0^\pi (\bar{y}y'' - y\bar{y}'')\,dz + (\lambda - \bar{\lambda})\int_0^\pi y\bar{y}\,dz = 0. \tag{15}$$

The first integral can be evaluated explicity:

$$\begin{aligned}\int_0^\pi (\bar{y}y'' - y\bar{y}'')\,dz &= \bar{y}y' - y\bar{y}' \,\Big|_0^\pi \\ &= [\bar{y}(\pi)\,y'(\pi) - y(\pi)\,\bar{y}'(\pi)] - [\bar{y}(0)\,y'(0) - y(0)\,\bar{y}'(0)].\end{aligned} \tag{16}$$

From (11) we see that

$$\begin{aligned} y'(0) &= -a_0 k, & \bar{y}'(0) &= -a_0\bar{k}, \\ y(0) &= a_1 k, & \bar{y}(0) &= a_1\bar{k}, \end{aligned}$$

so that

$$\bar{y}(0)\,y'(0) - y(0)\,\bar{y}'(0) = -a_1 a_0 k\bar{k} + a_1 a_0 k\bar{k} = 0,$$

and similarly at the end point π.

It follows that

$$\int_0^\pi (\bar{y}y'' - y\bar{y}'')\,dz = 0,$$

and a check shows that the same result holds for the boundary conditions (12). Next we observe that

$$\int_0^\pi y\bar{y}\,dz = \int_0^\pi |\,y\,|^2\,dz > 0,$$

so that (15) reduces to

$$\lambda - \bar{\lambda} = 0,$$

which clearly shows that λ is real, since no nonreal number can equal its complex conjugate. From (14) one now sees by inspection that the zeros for that case are given by

$$\begin{aligned} \tfrac{1}{2}\sqrt{\lambda} &= n, \qquad n = 0, 1, 2, \cdots, \\ \lambda &= 4n^2. \end{aligned}$$

The zeros of (13) can generally not be determined explicity, but certain conclusions can be drawn regarding their nature. But several possibilities must be considered. First we assume that $a_1 b_1 \neq 0$. Then (13) can be rewritten in the form

$$\sin \sqrt{\lambda}\,\pi = \frac{(a_1 b_0 - a_0 b_1)\sqrt{\lambda}\cos\sqrt{\lambda}\,\pi - a_0 b_0 \sin\sqrt{\lambda}\,\pi}{a_1 b_1 \lambda}.$$

For large λ the right side is small, and one then has

$$\sin\sqrt{\lambda}\,\pi \approx 0,$$

so that

$$\sqrt{\lambda} \approx n, \qquad n \text{ integral.}$$

The symbol $\approx$ should be read as approximately equal.

When $a_1 b_1 = 0$, but $a_1 b_0 - a_0 b_1 \neq 0$, (13) can be rewritten as

$$\cos\sqrt{\lambda}\,\pi = \frac{a_0 b_0 \sin\sqrt{\lambda}\,\pi}{(a_1 b_0 - a_0 b_1)\sqrt{\lambda}}.$$

For large λ the right side is small, so that

$$\cos\sqrt{\lambda}\pi \approx 0$$

and

$$\sqrt{\lambda} \approx n + \frac{1}{2}, \qquad n \text{ integral.}$$

Finally, when $a_1 b_1 = a_1 b_0 - a_0 b_1 = 0$, but $a_0 b_0 \neq 0$, we find from (13) that

$$\sin\sqrt{\lambda}\,\pi = 0,$$

so that

$$\sqrt{\lambda} = n, \qquad n = 0, 1, \cdots.$$

We have now shown that in the case of boundary conditions (11) and (12), one obtains infinite sets of real λ's corresponding to which (5) has nonvanishing solutions satisfying these boundary conditions. Evidently these λ's approach ∞, but it can be shown that they are bounded below, so that there must be a smallest λ. To show this, we suppose that λ is negative and sufficiently small, say,

$$\lambda = -\mu^2,$$

where μ is large, and seek a solution in the form

$$y = c_1 e^{\mu z} + c_2 e^{-\mu z}$$

satisfying the boundary conditions (11). Then

$$\Delta(\lambda) = (a_0 + a_1\mu)(b_0 - b_1\mu)\,e^{-\mu\pi} + (a_0 - a_1\mu)(b_0 + b_1\mu)\,e^{\mu\pi},$$

and for large μ we see that $\Delta(\lambda)$ has to grow like $e^{\mu\pi}$. Hence $\Delta(\lambda)$ cannot vanish for sufficiently large values of μ, or correspondingly, for sufficiently small values of λ. Boundary conditions (12) can be treated in a similar fashion.

One interesting difference between (13) and (14) is that (13) has no double zeros, whereas (14) does. This implies that (5) can have only one linearly independent solution satisfying (11), but may have two distinct solutions satisfying (12). First consider (12) and let $\lambda = 4n^2$. Then we find that

$$y = c_1 \sin 2nz + c_2 \cos 2nz, \qquad n \geqslant 1,$$
$$y = c, \qquad n = 0.$$

satisfies (5) and (12), and for $n \geqslant 1$, c_1 and c_2 are both arbitrary. For $n = 0$, we have only one arbitrary constant.

The most general solution of (5) satisfying (11) contains only one arbitrary constant, for if y is a solution, so clearly is cy, where c is arbitrary. Suppose that y_1 and y_2 satisfy (5) and (11); then

$$y = c_1 y_1 + c_2 y_2$$

also satisfies (5) and (11). Now we consider two separate cases, $a_1 = 0$ and $a_1 \neq 0$. In the former case we have from (11), since $a_0 \neq 0$,

$$y_1(0) = y_2(0) = 0,$$

so that

$$y(0) = c_1 y_1(0) + c_2 y_2(0) = 0.$$

We now choose c_1 and c_2 so that

$$y'(0) = c_1 y_1'(0) + c_2 y_1'(0) = 0.$$

This must be possible because both $y_1'(0)$, $i = 1, 2$, cannot vanish. Otherwise y_1 and y_2 would vanish identically. Then

$$y'(0) = 0,$$

and therefore y vanishes identically. Therefore y_1 and y_2 are linearly dependent.

In the second alternative where $a_1 \neq 0$, we find that neither $y_i(0)$, $i = 1, 2$, can vanish; otherwise y_i would vanish identically. Then we select c_1 and c_2 so that

$$y(0) = 0.$$

We then find that

$$y'(0) = c_1 y_1'(0) + c_2 y_2'(0) = \frac{-a_0}{a_1} c_1 y_1(0) - \frac{a_0}{a_1} c_2 y_2(0) = \frac{-a_0}{a_1} y(0) = 0.$$

Therefore we again find that y_1 and y_2 are linearly dependent. Hence the most general solution satisfying (11) contains only one arbitrary element.

The conclusions of the preceding discussion will now be summarized in the following theorem.

THEOREM: The differential equation

$$y'' + \lambda y = 0$$

with the separated boundary conditions

$$a_0 y(0) + a_1 y'(0) = 0,$$
$$b_0 y(\pi) + b_1 y'(\pi) = 0,$$

has no nonvanishing solution unless λ is a root of the equation

$$(a_1 b_0 - a_0 b_1)\sqrt{\lambda}\cos\sqrt{\lambda}\,\pi - (a_0 b_0 + a_1 b_1 \lambda)\sin\sqrt{\lambda}\,\pi = 0.$$

All roots of this equation are simple and can be arranged in an increasing sequence

$$\lambda_0 < \lambda_1 < \lambda_2 < \cdots,$$

the elements of which become arbitrarily large. Corresponding to each of these there exists a solution of the form

$$y_n = c_n\,[a_0 \sin\sqrt{\lambda_n}\,z - a_1\sqrt{\lambda_n}\cos\sqrt{\lambda_n}\,z], \qquad n = 0, 1, 2, \cdots,$$

where c_n is arbitrary. All these satisfy the integral relationship

$$\int_0^\pi y_n y_m\,dz = 0, \qquad n \neq m.$$

Corresponding to the periodicity boundary conditions

$$y(0) - y(\pi) = 0,$$
$$y'(0) - y'(\pi) = 0,$$

we find nonvanishing solutions only if λ is a root of

$$-4\sqrt{\lambda}\sin^2\frac{1}{2}\sqrt{\lambda}\,\pi = 0.$$

All roots can be arranged in an increasing sequence, where all double roots occur twice:

$$\lambda_0 < \lambda_1 = \lambda_2 < \lambda_3 = \lambda_4 < \cdots < \lambda_{2n+1} = \lambda_{2n} < \cdots.$$

We find corresponding solutions:

$$\begin{aligned} y_{2n} &= c_n \cos 2nz, \qquad n = 0, 1, 2, \cdots, \\ y_{2n+1} &= c_{2n+1} \sin 2nz. \end{aligned}$$

These also satisfy the integral relationship

$$\int_0^{\pi} y_n y_m \, dz = 0, \qquad n \neq m.$$

The only items not proved in the prior discussion are the integral relationships. We shall prove them for the case of separated boundary conditions. The proof in the case of the periodicity boundary conditions is similar. Thus we suppose that

$$\begin{aligned} y_n'' + \lambda_n y_n &= 0, \\ y_m'' + \lambda_m y_m &= 0, \end{aligned}$$

so that

$$y_m(y_n'' + \lambda_n y_n) - y_n(y_m'' + \lambda_m y) = y_m y_n'' - y_n y_m'' + (\lambda_n - \lambda_m)\, y_n y_m = 0.$$

Then

$$\int_0^{\pi} (y_m y_n'' - y_n y_m'')\, dz + (\lambda_n - \lambda_m) \int_0^{\pi} y_n y_m \, dz = 0,$$

but

$$\lambda_n - \lambda_m \neq 0, \qquad \text{for } n \neq m,$$

and

$$\int_0^{\pi} (y_m y_n'' - y_n y_m'')\, dz = 0$$

as in (16). It follows that

$$\int_0^{\pi} y_n y_m \, dz = 0.$$

4.4 THE EQUATION $y'' + [\lambda - q(z)]\, y = 0$

This section will be devoted to a discussion of the differential equation

$$y'' + [\lambda - q(z)]\, y = 0, \tag{1}$$

subject to the boundary conditions

$$\begin{aligned} a_0 y(0) + a_1 y'(0) &= 0, \\ b_0 y(\pi) + b_1 y'(\pi) &= 0. \end{aligned} \tag{2}$$

It will be assumed that the function $q(z)$ is continuous in the interval $0 \leqslant z \leqslant \pi$ and therefore is bounded as well, say,

$$| q(z) | \leqslant M.$$

The asymptotic nature of (1) for large values of λ was discussed in Example 4, Section 3.4. There it was shown that one can find two linearly independent solutions, whose asymptotic forms are

$$V_1 = e^{-i\sqrt{\lambda}z} \left[\frac{1}{2} + \frac{i}{4}\int q\,dz\,\lambda^{-1/2} + \cdots\right],$$

$$V_2 = e^{i\sqrt{\lambda}z} \left[\frac{1}{2} - \frac{i}{4}\int q\,dz\,\lambda^{-1/2} + \cdots\right].$$

From these one can construct two other solutions:

$$\begin{aligned} y_1 &= V_1 + V_2 = \cos\sqrt{\lambda}\,z + \frac{1}{2\sqrt{\lambda}} \sin\sqrt{\lambda}\,z \int q\,dz + \cdots, \\ y_2 &= i(V_1 - V_2) = \sin\sqrt{\lambda}\,z - \frac{1}{2\sqrt{\lambda}} \cos\sqrt{\lambda}\,z \int q\,dz + \cdots. \end{aligned} \tag{3}$$

From these asymptotic solutions it seems plausible that for large λ, the conclusions reached in the preceding section for the case $q(z) = 0$ should hold true. But in view of the fact that the convergence of these series is not assured, a somewhat different and rigorous approach will be used.

We first consider the inhomogeneous equation

$$y'' + \lambda y = f(z) \tag{4}$$

and solve it by the method of variation of parameters. To do so, we rewrite (4) as a system. Let

$$x_1 = y,$$
$$x_2 = y',$$

so that

$$X' = \begin{pmatrix} 0 & 1 \\ -\lambda & 0 \end{pmatrix} X + \begin{pmatrix} 0 \\ f(z) \end{pmatrix}.$$

The matrix

$$\Phi = \begin{pmatrix} \cos\sqrt{\lambda}\,z & \sin\sqrt{\lambda}\,z \\ -\sqrt{\lambda}\sin\sqrt{\lambda}\,z & \sqrt{\lambda}\cos\sqrt{\lambda}\,z \end{pmatrix}$$

satisfies

$$\Phi' = \begin{pmatrix} 0 & 1 \\ -\lambda & 0 \end{pmatrix} \Phi.$$

Now the new independent variable Y is introduced by

$$X = \Phi Y,$$

so that

$$Y' = \Phi^{-1}\begin{pmatrix} 0 \\ f(z) \end{pmatrix} = \begin{pmatrix} \cos\sqrt{\lambda}\, z & \dfrac{-\sin\sqrt{\lambda}\, z}{\sqrt{\lambda}} \\ \sin\sqrt{\lambda}\, z & \dfrac{\cos\sqrt{\lambda}\, z}{\sqrt{\lambda}} \end{pmatrix}\begin{pmatrix} 0 \\ f(z) \end{pmatrix} = \begin{pmatrix} \dfrac{-\sin\sqrt{\lambda}\, z}{\sqrt{\lambda}} f(z) \\ \dfrac{\cos\sqrt{\lambda}\, z}{\sqrt{\lambda}} f(z) \end{pmatrix}.$$

From this one can now obtain Y by an integration and then X, so that

$$X = \Phi\left[\begin{pmatrix} c_1 \\ c_2 \end{pmatrix} + \int_0^z \begin{pmatrix} \dfrac{-\sin\sqrt{\lambda}\, t\, f(t)}{\sqrt{\lambda}} \\ \dfrac{\cos\sqrt{\lambda}\, t}{\sqrt{\lambda}} f(t) \end{pmatrix} dt\right],$$

where c_1 an c_2 are arbitrary constants. Then one finally obtains

$$y = c_1 \cos\sqrt{\lambda}\, z + c_2 \sin\sqrt{\lambda} z + \int_0^z f(t)\, \frac{\sin\sqrt{\lambda}\,(z - t)}{\sqrt{\lambda}}\, dt. \tag{5}$$

Now (4) reduces to (1) if one replaces $f(z)$ by $q(z)\, y$. In this case (5) reduces to

$$y = c_1 \cos\sqrt{\lambda}\, z + c_2 \sin\sqrt{\lambda}\, z + \int_0^z q(t)\, \frac{\sin\sqrt{\lambda}\,(z - t)}{\sqrt{\lambda}}\, y(t)\, dt, \tag{6}$$

and (6) is an integral equation for y, which is equivalent to the differential equation (1). By differentiation (1) can be recovered from (6). This type of equation is known as a Volterra integral equation. The coefficients c_1 and c_2 can be determined from the boundary conditions (2). From the general existence theorems in Chapter 2 it is known that if the coefficients of a linear differential equation are continuous, so is the solution. Therefore the solution y must be continuous for all $0 \leqslant z \leqslant \pi$, and hence bounded so that a positive K exists such that

$$|y| \leqslant K, \qquad \text{for } 0 \leqslant z \leqslant \pi.$$

Then the integral in (6) can be estimated so that

$$\begin{aligned} \left| \int_0^z q(t)\, \frac{\sin\sqrt{\lambda}\,(z - t)\, y}{\sqrt{\lambda}}\, y(t)\, dt \right| & \\ &\leqslant \int_0^z |q(t)| \left| \frac{\sin\sqrt{\lambda}\,(z - t)}{\sqrt{\lambda}} \right| |y(t)|\, dt \\ &\leqslant \frac{1}{\sqrt{\lambda}} MK\pi. \end{aligned} \tag{7}$$

It follows that

$$y = c_1 \sin \sqrt{\lambda}\, z + c_2 \cos \sqrt{\lambda}\, z + 0\left(\frac{1}{\sqrt{\lambda}}\right) \tag{8}$$

The symbol $0(1/\sqrt{\lambda})$ represents a function that has the property that $\sqrt{\lambda}\, 0(1/\sqrt{\lambda})$ is bounded for large λ. Therefore $0(1/\sqrt{\lambda})$ must become arbitrarily small as λ becomes sufficiently large. This conclusion was intuitively evident from (3), but has now been rigorously established.

From the preceding discussion it is clear that the asymptotic solutions (3) can be rewritten as

$$\begin{aligned} y_1 &= \cos \sqrt{\lambda}\, z + R_1(z, \lambda), \\ y_2 &= \sin \sqrt{\lambda}\, z + R_2(z, \lambda), \end{aligned} \tag{9}$$

where the $R_i(z, \lambda)$ are functions of z and λ such that $\sqrt{\lambda}\, R_i(z, \lambda)$ is bounded for all λ. By differentiating (6) with respect to z and examining the integral, one can show as in (7) that the remainder terms $R_i'(z, \lambda)$ are bounded. By using (9), (8) can now be rewritten as

$$y = c_1 y_1 + c_2 y_2.$$

To determine c_1 and c_2, we must make use of the boundary conditions (2) and find

$$[a_0 y_1(0) + a_1 y_1'(0)]\, c_1 + [a_0 y_2(0) + a_1 y_2'(0)]\, c_2 = 0,$$

$$[b_0 y_1(\pi) + b_1 y_1'(\pi)]\, c_1 + [b_0 y_2(\pi) + b_1 y_2'(\pi)]\, c_2 = 0.$$

In order for this homogeneous system to have a solution, we require that its determinant should vanish. Then, using (9), we find

$$\Delta(\lambda) = \begin{vmatrix} a_0(1 + R_1(0, \lambda)) + a_1 R_1'(0, \lambda) & a_0 R_2(0, \lambda) + a_1(\sqrt{\lambda} + R_2'(0, \lambda)) \\ b_0(\cos \sqrt{\lambda}\, \pi + R_1(\pi, \lambda)) & b_0(\sin \sqrt{\lambda}\, \pi + R_2(\pi, \lambda)) \\ \quad + b_1(-\sqrt{\lambda} \sin \sqrt{\lambda}\, \pi & \quad + b_1(\sqrt{\lambda} \cos \sqrt{\lambda}\, \pi \\ \quad + R_1'(\pi, \lambda)) & \quad + R_2'(\pi, \lambda)) \end{vmatrix} \tag{10}$$

When $a_1 b_1 \neq 0$, a calculation shows that

$$\Delta(\lambda) = \lambda a_1 b_1 \sin \sqrt{\lambda}\, \pi + \sqrt{\lambda}\, P_1(\lambda),$$

where $P_1(\lambda)$ is bounded for all λ. To find the zeros of $\Delta(\lambda)$, we require that

$$\sin \sqrt{\lambda}\, \pi = \frac{-P_1(\lambda)}{\sqrt{\lambda}\, a_1 b_1},$$

and for large λ the right side is small so that

$$\sqrt{\lambda} \approx n, \qquad n \text{ integral.}$$

When $a_1 b_1 = 0$, but $a_0 b_1 - a_1 b_0 \neq 0$,

$$\Delta(\lambda) = (a_0 b_1 - a_1 b_0) \sqrt{\lambda} \cos \sqrt{\lambda}\, \pi + P_2(\lambda),$$

where $P_2(\lambda)$ is bounded for all λ. Then

$$\cos \sqrt{\lambda}\, \pi = \frac{-P_2(\lambda)}{(a_0 b_1 - a_1 b_0) \sqrt{\lambda}}$$

and

$$\sqrt{\lambda} \approx n + \frac{1}{2}, \qquad n \text{ integral.}$$

When $a_1 b_1 = a_0 b_1 - a_1 b_0 = 0$, we have

$$\Delta(\lambda) = a_0 b_0 \sin \sqrt{\lambda}\, \pi + \frac{1}{\sqrt{\lambda}} P_3(\lambda)$$

and $P_3(\lambda)$ is bounded, we find as before that the zeros are such that

$$\sqrt{\lambda} \approx n, \qquad n \text{ integral.}$$

Hence, as in the preceding section, we find that (1) can have solutions satisfying (2) only if the λ's are zeros of $\Delta(\lambda)$. These zeros must have a lower bound, since for sufficiently small λ, the solutions have an exponential character and cannot satisfy both boundary conditions (2). The argument is almost identical to the one presented in the preceding section. There it was also shown that in the case of separated boundary conditions, all zeros of $\Delta(\lambda)$ are simple. The argument there was based on the idea that two solutions satisfying the same end-point condition,

$$a_0 y(0) + a_1 y'(0) = 0,$$

must be linearly dependent. By the same argument it follows that here also all zeros of $\Delta(\lambda)$ are simple. Thus we find that there exists an infinite sequence

$$\lambda_0 < \lambda_1 < \lambda_2 < \cdots$$

and a corresponding set of functions $y_0, y_1, y_2, \cdots$ satisfying

$$y_n'' + [\lambda_n - q(z)]\, y_n = 0$$

and boundary conditions (2). As before we can show that

$$\int_0^\pi y_n y_m \, dz = 0, \qquad n \neq m,$$

$$y_n[y_m'' + (\lambda_m - q(z))\, y_m] - y_m[y_n'' + (\lambda_n - q(z)\, y_n] = 0,$$

so that

$$y_n y_m'' - y_m y_n'' + (\lambda_m - \lambda_n)\, y_n y_m = 0. \tag{11}$$

From the fact that

$$\int_0^\pi (y_n y_m'' - y_m y_n'')\, dz = 0,$$

$$\lambda_m - \lambda_n \neq 0,$$

we see that

$$\int_0^\pi y_n y_m \, dz = 0, \qquad \text{for } n \neq m.$$

The following theorem has therefore been proved.

THEOREM: The differential equation

$$y'' + [\lambda - q(z)]\, y = 0$$

with the boundary conditions

$$a_0 y(0) + a_1 y'(0) = 0,$$
$$b_0 y(\pi) + b_1 y'(\pi) = 0,$$

where $q(z)$ is a continuous and bounded function for $0 \leqslant z \leqslant \pi$, can have nonvanishing solutions if and only if λ belongs to an infinite sequence

$$\lambda_0 < \lambda_1 < \lambda_2 < \cdots.$$

This sequence has no finite limit point. Corresponding to each of these, there are solutions $y_0, y_1, y_2, \cdots$, and they satisfy the integral relationships

$$\int_0^\pi y_n y_m \, dz = 0, \qquad n \neq m.$$

Next the following theorem will be proved.

THEOREM: The solution $y_n(z)$ in the preceding theorem has precisely n simple zeros in the interval $(0, \pi)$.

If z_1 were a multiple zero, it would follow that $y(z_1) = y'(z_1) = 0$. But a homogeneous second-order differential equation with these initial conditions has no nonvanishing solution. Hence all zeros are simple. The remainder of the proof will be conducted in several stages. First we shall show that if $y_n(z)$ has k zeros in $(0, \pi)$, then $y_{n+1}(z)$ will have $k + 1$ zeros. Suppose $y_n(z)$ vanishes at z_1 and z_2 and $y_n(z) > 0$ in (z_1, z_2); then $y_{n+1}(z)$ must vanish at least once in (z_1, z_2). Otherwise we can assume that $y_{n+1}(z) > 0$ in (z_1, z_2). Then we must have

$$\int_{z_1}^{z_2} y_n(z)\, y_{n+1}(z)\, dz > 0, \tag{12}$$

since the integral of a positive function must be positive.

From the nature of y_n we see that

$$y_n'(z_1) > 0, \qquad y_n'(z_2) < 0,$$

and also

$$y_{n+1}(z_1) > 0, \qquad y_{n+1}(z_2) > 0.$$

From (11) we find, for $m = n + 1$, since

$$\lambda_{n+1} - \lambda_n > 0$$

and

$$\int_{z_1}^{z_2} (y_n y_{n+1}'' - y_{n+1} y_n'')\, dt = y_n y_{n+1}' - y_{n+1} y_n' \Big|_{z_1}^{z_2}$$

$$= y_{n+1}(z_1)\, y_n'(z_1) - y_{n+1}(z_2)\, y_n'(z_2) > 0,$$

that

$$\int_{z_1}^{z_2} y_n y_{n+1}\, dz < 0,$$

contradicting (12).

Therefore y_{n+1} must vanish at least once in (z_1, z_2). Similarly one can show that between any two zeros of y_{n+1}, y_n must vanish at least once. Therefore, between any two consecutive zeros of y_n, y_{n+1} vanishes precisely once.

If z_1 denotes the first zero of y_n greater than 0, y_{n+1} must vanish at least once in $(0, z_1)$. The proof is as before. We suppose that

$$a_0 y_n(0) + a_1 y_n'(0) = 0, \qquad y_n(0) > 0; \quad y_n'(z_1) < 0,$$

and

$$a_0 y_{n+1}(0) + a_1 y_{n+1}'(0) = 0, \qquad y_{n+1}(z) > 0 \qquad \text{in } (0, z_1).$$

Then

$$\int_0^{z_1} (y_n y_{n+1}'' - y_{n+1} y_n'')\, dz = y_n y_{n+1}' - y_{n+1} y_n' \Big|_0^{z_1} = -y_{n+1}(z_1)\, y_n'(z_1) > 0,$$

since the contribution from $z = 0$ vanishes by virtue of the boundary conditions. Therefore

$$\int_0^{z_1} y_n y_{n+1} \, dz < 0,$$

again contradicting (12). Similarly, if $y_{n+1}(z_1) = 0$, it would follow that y_n would vanish in the interval $(0, z_1)$.

It follows, therefore, that if $y_n(z)$ has k zeros such that

$$0 < z_1 < z_2 < \cdots < z_k < \pi,$$

then $y_{n+1}(z)$ must have $k + 1$ zeros $z_1', z_2', \cdots z_{k+1}'$ such that

$$0 < z_1' < z_1 < z_2' < z_2 < \cdots < z_k < z_{k+1}' < \pi.$$

It remains to show that $k = n$. To do so, all that needs to be proved is that y_0 has no zero in $(0, \pi)$. We suppose that z_1 is the first zero of $y_0(z)$ greater than 0. Let y be any solution of

$$y'' + (\lambda - q(z))\, y = 0$$

satisfying the condition

$$a_0 y(0) + a_1 y'(0) = 0.$$

We shall now use the fact, without proof, that the zeros of y are continuous functions of λ. From the preceding discussion it follows that if $\lambda > \lambda_0$, y will vanish in $(0, z_1)$. Let $z(\lambda)$ denote that zero of y which, for $\lambda = \lambda_0$, reduces to z_1. Evidently $z(\lambda)$ is a decreasing function of λ. Therefore for some sufficiently small λ, say, λ^*, we shall have $z(\lambda^*) = \pi$. Then y satisfies the system

$$y'' + [\lambda^* - q(z)]\, y = 0,$$
$$a_0 y(0) + a_1 y'(0) = 0,$$
$$y(\pi) = 0,$$

and $y > 0$ for $0 \leqslant z < \pi$.

In the case where b_1 in the boundary conditions at π vanishes, we have found a new λ^* that is smaller than λ_0, and a solution y, satisfying the boundary condition, which does not vanish in $(0, \pi)$.

When $b_1 \neq 0$, we have

$$b_0 y(\pi) + b_1 y'(\pi) = b_1 y'(\pi) \neq 0, \tag{13}$$

since $y'(\pi) < 0$, by the nature of y.

If λ becomes sufficiently small, we see that y must have a typical exponential behavior. The following argument is of an asymptotic nature, but can as in previous discussions be made rigorous. Then

$$y = e^{\sqrt{-\lambda} z} + \cdots,$$

so that

$$b_0 y(\pi) + b_1 y'(\pi) = \sqrt{-\lambda}\, b_1 \, e^{\sqrt{\lambda}\pi} + \cdots. \tag{14}$$

The right sides of (13) and (14) are opposite in sign by hypothesis. It follows therefore that $b_0 y(\pi) + b_1 y'(\pi)$ changes sign as λ decreases sufficiently from λ^*. It is a continuous function of λ and therefore must change sign for some value of λ smaller than λ^*. Hence we can find a new λ_0 such that y satisfies (1) and (2) and does not vanish in $(0, \pi)$. Therefore $y_0(z)$ has no zeros in $(0, \pi)$, and generally $y_n(z)$ has precisely n zeros in $(0, \pi)$.

From now on we shall refer to the λ_n as eigenvalues and the y_n as the eigenfunctions of the boundary value problem (1), (2). This type of problem is often referred to in the literature as a Sturm-Liouville problem, and the set of all λ_n is called the Sturm-Liouville spectrum. In many physical problems, which in their mathematical formulation lead to Sturm-Liouville problem, one can determine the spectrum experimentally. Often the function $q(z)$ in (1) is unknown, and the question arises to what extent $q(z)$ can be reconstructed. This type of problem is known as an inverse Sturm-Liouville spectrum. We shall not attempt to deal with these problems here, but a few results derived will provide some information.

First we shall consider the simple problem

$$\begin{gathered} y'' + \mu y = 0, \\ y(0) = y(\pi) = 0. \end{gathered} \tag{15}$$

We find that the general solution is given by

$$y = c_1 \sin \sqrt{\mu}\, z + c_2 \cos \sqrt{\mu}\, z.$$

From

$$y(0) = c_2$$

we require that

$$c_2 = 0,$$

and then

$$y(\pi) = c_1 \sin \sqrt{\mu}\, \pi = 0.$$

In order to find nonvanishing solutions, we require that

$$\sqrt{\mu_n} = n, \qquad n = 1, 2, 3, \cdots.$$

Then the eigenfunctions are given by

$$y_n = c_n \sin nz.$$

Now we turn to the more general case:

$$\begin{gathered} y'' + [\lambda - q(z)]\, y = 0, \\ y(0) = y(\pi) = 0. \end{gathered} \tag{16}$$

We shall assume that λ is so large that $\lambda - q(z) > 0$ and that $q(z)$ is differentiable It is plausible that the spectrum of (16) will be approximated by that of (15) for large λ. We shall suppose that we can find a solution in the form

$$y = A(z) \sin \phi(z),$$
$$y' = \sqrt{\lambda - q(z)}\, A(z) \cos \phi(z). \tag{17}$$

These are suggested by the solution of (15), since there $q(z) = 0$, A is a constant and $\phi(z) = \sqrt{\lambda}\, z$, and y and y' are as in (17). We shall now derive new differential equations to determine $A(z)$ and $\phi(z)$. If we differentiate the first of the above equations and equate the result to the second, and differentiate the second equation and use (16), we obtain

$$A' \sin \phi + A\phi' \cos \phi = \sqrt{\lambda - q(z)}\, A \cos \phi$$
$$\sqrt{\lambda - q}\, A' \cos \phi - \sqrt{\lambda - q}\, A\phi' \sin \phi - \frac{q'}{2\sqrt{\lambda - q}} A \cos \phi = -[\lambda - q]\, A \sin \phi.$$

By eliminating A' from the above, one obtains the two simultaneous first-order equations

$$\phi' = \sqrt{\lambda - q} - \frac{1}{4} \frac{q'}{\lambda - q} \sin 2\phi,$$
$$A' = A \frac{q'}{2(\lambda - q)} \cos^2 \phi. \tag{18}$$

The first of these is independent of A and, once ϕ has been determined, A is found from the second equation by an integration. To satisfy the boundary conditions in (16), we impose on ϕ the conditions

$$\phi(0) = 0, \qquad \phi(\pi) = n\pi. \tag{19}$$

The first of the equations (18) is nonlinear. In a later chapter existence proofs will be furnished. For the present, we shall assume all necessary facts. For a first-order equation we require only one initial condition to specify a solution uniquely. The second condition of (19) will not necessarily be satisfied unless λ is an eigenvalue. This condition will be used to determine the required value of λ. By use of the first equation of (18) and the condition at $z = 0$, one finds

$$\phi = \int_0^z \sqrt{\lambda - q}\, dz - \frac{1}{4} \int_0^z \frac{q'}{\lambda - q} \sin 2\phi\, dz, \tag{20}$$

which is a Volterra integral equation for ϕ. Under the assumption that q' and q are bounded and that λ is sufficiently large, one finds that

$$\left| \int_0^z \frac{q'}{\lambda - q} \sin 2\phi\, dz \right| < \frac{K}{\lambda}$$

for some constant K. Then (20) can be rewritten as

$$\phi = \int_0^z \sqrt{\lambda - q}\, dz + 0\left(\frac{1}{\lambda}\right). \tag{21}$$

[The notation $0(1/\lambda)$, as explained before, indicates that the term $0(1/\lambda)$ is such that $\lambda 0(1/\lambda)$ is bounded for large λ.]

Therefore, by using (21), we find

$$\sin 2\phi = \left(\sin 2 \int_0^z \sqrt{\lambda - q}\, dz\right) \cos 0\left(\frac{1}{\lambda}\right) + \left(\cos 2 \int_0^z \sqrt{\lambda - q}\, dz\right) \sin 0\left(\frac{1}{\lambda}\right).$$

For large λ we have by the mean value theorem,

$$\cos 0\left(\frac{1}{\lambda}\right) = 1 + 0\left(\frac{1}{\lambda^2}\right),$$

$$\sin 0\left(\frac{1}{\lambda}\right) = 0\left(\frac{1}{\lambda}\right),$$

so that

$$\sin 2\phi = \sin 2 \int_0^z \sqrt{\lambda - q}\, dz + 0\left(\frac{1}{\lambda}\right). \tag{22}$$

After (22) is inserted into (20), one obtains

$$\phi = \int_0^z \sqrt{\lambda - q}\, dz - \frac{1}{4}\int_0^z \frac{q'}{\lambda - q} \sin\left(2\int_0^z \sqrt{\lambda - q}\, dz\right) dz + 0\left(\frac{1}{\lambda^2}\right). \tag{23}$$

The interesting thing about (23) is that for terms up to order $1/\lambda^2$, the right side is known, so that ϕ is known approximately. In particular for $z = \pi$, we have from (19),

$$n\pi = \phi(\pi) = \int_0^\pi \sqrt{\lambda - q}\, dz - \frac{1}{4}\int_0^\pi \frac{q'}{\lambda - q} \sin\left(2\int_0^z \sqrt{\lambda - q}\, dz\right) dz + 0\left(\frac{1}{\lambda^2}\right). \tag{24}$$

The preceding equation will yield estimates for λ. It will be convenient to assume at this point that q is triply differentiable, in order to be able to perform an integration by parts, and that q has mean value 0; that is,

$$\int_0^\pi q\, dz = 0.$$

The latter is no restriction, since if

$$\int_0^\pi q\, dz = m,$$

we write

$$\sqrt{\lambda - q} = \sqrt{(\lambda - m) - (q - m)},$$

where $\lambda - m$ is a new parameter and $q - m$ has mean value 0.

Then the first term on the right of (24) becomes by the binomial theorem,

$$\int_0^\pi \sqrt{\lambda - q}\, dt = \int_0^\pi \left[\sqrt{\lambda} - \frac{q}{2\sqrt{\lambda}} - \frac{q^2}{8\lambda^{3/2}} + 0\left(\frac{1}{\lambda^{5/2}}\right)\right] dz$$

$$= \sqrt{\lambda}\,\pi - \frac{1}{8\lambda^{3/2}} \int_0^\pi q^2\, dz + 0\left(\frac{1}{\lambda^{5/2}}\right). \tag{25}$$

We can apply an integration by parts to the second term on the right of (24) and find

$$\int_0^\pi \frac{q'}{\lambda - q} \sin\left(2\int_0^z \sqrt{\lambda - q}\, dz\right) dz = \frac{-q'}{2(\lambda - q)^{3/2}} \cos 2 \int_0^z \sqrt{\lambda - q}\, dz \,\Big|_0^\pi$$

$$+ \int_0^\pi \left(\frac{d}{dz}\frac{q'}{2(\lambda - q)^{3/2}}\right) \cos\left(2\int_0^z \sqrt{\lambda - q}\, dz\right) dz$$

$$= \frac{-q'(\pi)\cos\left(2\int_0^\pi \sqrt{\lambda - q}\, dz\right)}{2(\lambda - q(\pi))^{3/2}} + \frac{q'(0)}{2(\lambda - q(0))^{3/2}} + 0\left(\frac{1}{\lambda^2}\right). \tag{26}$$

The fact that the integrated term in (26) is $0(1/\lambda^2)$ follows from a second integeration by parts. By using the estimates in (24) and (25), one can now reduce (26) further to obtain

$$\int_0^\pi \frac{q'}{\lambda - q} \sin\left(2\int_0^z \sqrt{\lambda - q}\, dz\right) dz = \frac{-q'(\pi) + q'(0)}{2\lambda^{3/2}} + 0\left(\frac{1}{\lambda^2}\right). \tag{27}$$

By use of (25) and (27), (24) becomes

$$n\pi = \sqrt{\lambda}\,\pi - \frac{\int_0^\pi q^2\, dz - q'(\pi) + q'(0)}{8\lambda^{3/2}} + 0\left(\frac{1}{\lambda^2}\right).$$

From this equation one can determine an asymptotic series for $\sqrt{\lambda}$ in terms of n. Then one finds

$$\sqrt{\lambda_n} = n + \frac{\int_0^\pi q^2\, dz - q'(\pi) + q'(0)}{8n^3\pi} + 0\left(\frac{1}{n^4}\right). \tag{28}$$

For $q = 0$ it was shown previously that

$$\sqrt{\mu_n} = n,$$

so that

$$\sqrt{\lambda_n} - \sqrt{\mu_n} = \frac{C}{n^3} + 0\left(\frac{1}{n^4}\right),$$

where C is defined in (28).

This last result was derived for problem (16), but by similar methods can be shown to hold for the general boundary conditions (2). We therefore have proved the following theorem.

THEOREM: The Sturm-Liouville spectrum $\{\lambda_n\}$ for

$$y'' + [\lambda - q(z)]y = 0$$

with the boundary conditions

$$a_0 y(0) + a_1 y'(0) = 0,$$
$$b_0 y(\pi) + b_1 y'(\pi) = 0,$$

where $q(z)$ is triply differentiable and has mean value 0, when compared to the spectrum $\{\mu_n\}$ for

$$y'' + \mu y = 0$$

with the same boundary conditions is such that

$$\sqrt{\lambda_n} - \sqrt{\mu_n} = \frac{C}{n^3} + 0\left(\frac{1}{n^4}\right).$$

C depends only on $\int_0^\pi q^2\, dz$, $q'(\pi)$, and $q'(0)$.

Thus, if in a physical problem one can determine the λ_n experimentally, the numbers C in the above theorem can be determined, and some information regarding the nature of $q(z)$ has then been found.

In many important problems one encounters equations of type (1), where either $q(z)$ is not bounded in the interval $[0, \pi]$ or the interval under consideration is infinite. The general theory underlying these problems is quite complicated and will not be touched on here, but the following examples will cover some of the more important of these cases.

EXAMPLE 1. The Bessel equation

$$R'' + \frac{1}{r} R' + \lambda R = 0 \tag{29}$$

with the boundary conditions

$$R'(0) = 0, \qquad R(a) = 0,$$

was set up in Section 4.1 in connection with the problem of vibrations of a circular plate. The condition $R'(0) = 0$ was selected from symmetry considerations. In Chapter 3 it was seen that one solution could be formed as an infinite series in the form

$$R = \sum_0^\infty \frac{\left(\frac{i}{2}\sqrt{\lambda}\, r\right)^{2n}}{[n!]^2}. \tag{30}$$

This solution evidently satisfies the boundary condition

$$R'(0) = 0.$$

A second linearly independent solution was found, but it contained terms of the form $\ln r$. These were not regular at $r = 0$ and certainly cannot satisfy the condition $R'(0) = 0$. Hence the above series solution is the one that satisfies the boundary condition at $r = 0$. Evidently the condition that R should be finite at $r = 0$ would also serve to single out this solution. To determine λ, one uses the second boundary condition and finds

$$\sum_0^\infty \frac{\left(\frac{i}{2}\sqrt{\lambda}\, a\right)^{2n}}{[n!]^2} = 0. \tag{31}$$

This is a transcendental equation for λ, whose roots are the eigenvalues of the problem. From the asymptotic formulas developed in Chapter 3, it follows that the asymptotic form of R for large λr must be

$$R \approx \frac{A \sin(\sqrt{\lambda}\, r + \phi)}{\sqrt{r}}.$$

To obtain estimates for the large solutions of (31), one can replace (31) by

$$A \frac{\sin(\sqrt{\lambda}\, a + \phi)}{\sqrt{a}} \approx 0,$$

showing that

$$\sqrt{\lambda_n} = \frac{n\pi}{a} + 0(1)$$

where $0(1)$ denotes a function bounded for all n.

This problem could also be solved by converting (29) to the Liouville normal form, as was shown in Example 1, Section 4.2. The substitution

$$R = \frac{u}{\sqrt{r}}$$

reduces (29) to

$$u'' + \left[\lambda + \frac{1}{4r^2}\right] u = 0.$$

Example 2. The Legendre equation

$$(1 - z^2)y'' - 2zy' + \lambda y = 0 \tag{32}$$

has singularities at the end points of the interval $(-1, 1)$. The substitutions

$$z = \cos\theta,$$

$$y = \frac{u(\theta)}{(\sin\theta)^{1/2}},$$

reduce (32) to

$$u''(\theta) + \left[\lambda + \frac{1}{4} + \frac{1}{4\sin^2\theta}\right] u(\theta) = 0 \tag{33}$$

and the interval $(-1, 1)$ reduces to the interval $0 \leqslant \theta \leqslant \pi$. The term $1/(4\sin^2\theta)$ in (33) is singular at the end points of the interval. The boundary conditions to be applied are that u be regular at the end points.

The coefficients in (32) are regular for $|z| < 1$. Therefore it must be possible to find a solution in the form

$$y = \sum_{0}^{\infty} a_n z^n, \tag{34}$$

which converges for $|z| < 1$. If (34) is inserted into (32), one obtains, as in Chapter 3, the recurrence formula

$$a_{n+2} = \frac{n(n+1) - \lambda}{(n+2)(n+1)} a_n. \tag{35}$$

If λ is of the form $k(k+1)$, where k is an integer, one sees that

$$a_{k+2} = 0,$$

and all subsequent coefficients must vanish. In this case the solution must be a polynomial, which of course is regular at $|z| = 1$. The resultant polynomials are known as the Legendre polynomials. A calculation shows that for

$$\lambda_0 = 0, \qquad P_0(z) = 1$$

$$\lambda_1 = 1 \cdot 2 \qquad P_1(z) = z$$

$$\lambda_2 = 2 \cdot 3 \qquad P_2(z) = \frac{1}{2}(3z^2 - 1)$$

$$\lambda_3 = 3 \cdot 4 \qquad P_3(z) = \frac{1}{2}(5z^3 - 3z)$$

$$\lambda_4 = 4 \cdot 5 \qquad P_4(z) = \frac{1}{8}(35z^4 - 30z^2 + 3)$$

and more generally for $\lambda_k = k(k+1)$,

$$P_k(z) = \frac{(2k)!}{2^k k!^2}\left[z^k - \frac{k(k-1)}{2(2k-1)} z^{k-2} + \frac{k(k-1)(k-2)(k-3)}{2 \cdot 4 \cdot (2k-1)(2k-3)} z^{k-4} + \cdots\right]. \tag{36}$$

By the methods of Chapter 2 the second linearly independent solution of (32) can be shown to be unbounded at $z^2 = 1$.

To discuss the case where $\lambda \neq k(k+1)$, we compare (35) to the simpler recurrence equation

$$b_{n+2} = \frac{n(n+1)}{(n+2)(n+1)} b_n. \tag{37}$$

The solution of this equation is given by

$$b_{2n} = \frac{c_0}{2n}, \qquad n = 1, 2, \cdots,$$

$$b_{2n+1} = \frac{c_1}{2n+1}, \qquad n = 0, 1, 2, \cdots,$$

where c_0 and c_1 are arbitrary. Then the series

$$\begin{aligned} \sum_1^\infty b_n z^n &= \frac{c_0}{2} \sum_1^\infty \frac{z^{2n}}{n} + c_1 \sum_1^\infty \frac{z^{2n+1}}{2n+1} \\ &= \frac{-c_0}{2} \ln(1 - z^2) + \frac{c_1}{2} \ln \frac{1+z}{1-z}, \end{aligned} \tag{38}$$

has singularities at $z = \pm 1$. If it can be shown that the solutions of (35) and (37) are so similar that the convergence properties of (34) and (38) are similar, then (34) will also be singular at $z = \pm 1$. If we let

$$c_n = \ln \frac{a_n}{b_n}$$

and divide (35) by (37) and take logarithms, we obtain

$$c_{n+2} = c_n + \ln \left[1 - \frac{\lambda}{n(n+1)}\right].$$

This recurrence formula can be solved explicitly, and one finds that

$$\begin{aligned} c_n = \ln \left[1 - \frac{\lambda}{(n-1)(n-2)}\right] &+ \ln \left[1 - \frac{\lambda}{(n-3)(n-4)}\right] \\ &+ \ln \left[1 - \frac{\lambda}{(n-5)(n-6)}\right] + \cdots. \end{aligned}$$

Since for large n, one has by the mean value theorem,

$$\ln \left[1 - \frac{\lambda}{(n-1)(n-2)}\right] \approx - \frac{\lambda}{(n-1)(n-2)},$$

and the series

$$\sum_{1}^{\infty} \frac{1}{n(n+1)}$$

converges as n tends to infinity, c_n approaches a constant. Therefore

$$\lim_{n\to\infty} \frac{a_n}{b_n} = \lim_{n\to\infty} e^{c_n}.$$

It follows that for large n,

$$a_n \approx b_n$$

and (34) must diverge at $z = \pm 1$.

Therefore all eigenvalues and eigenfunctions of (32) corresponding to regularity conditions at $z^2 = 1$ are given by (36).

EXAMPLE 3. In the study of the linear harmonic oscillator in quantum mechanics one encounters the Hermite differential equation

$$y'' + [\lambda - z^2]y = 0 \tag{39}$$

over the interval $-\infty < z < \infty$, and subject to the condition

$$\int_{-\infty}^{\infty} y^2\, dz < \infty. \tag{40}$$

In order to satisfy (40) it will be necessary that y should decay sufficiently fast at infinity. To study (39), it is convenient to use the substitution

$$y = e^{-z^2/2}u \tag{41}$$

so that (39) becomes

$$u'' - 2zu' + [\lambda - 1]u = 0. \tag{42}$$

Next one seeks a series solution for u and lets

$$u = \sum_{0}^{\infty} a_n z^n. \tag{43}$$

Then the a_n must satisfy

$$a_{n+2} = \frac{2n + 1 - \lambda}{(n+2)(n+1)}\, a_n. \tag{44}$$

When $\lambda = 2k + 1$, it follows that $a_{k+2} = 0$, and all subsequent a_n must vanish. Then (43) is a polynomial and 41 satisfies (40).

For $\lambda \neq 2k + 1$, one can show, as in the preceding example, that for large n,

$$a_{2n} \approx \frac{c_0}{n!}, \qquad n = 0, 1, \cdots .$$

$$a_{2n+1} \approx \frac{c_1}{n!}$$

It then follows from (43) that

$$u \approx c_0 e^{z^2} + c_1 z(e^{z^2} - 1),$$

and inserting this result into (41), it follows that

$$y \approx e^{z^2/2},$$

which certainly cannot satisfy (40). Then one obtains the following eigenvalues and eigenfunctions for (39):

$$\begin{aligned} \lambda_0 &= 1, & H_0(z) &= e^{-z^2/2}, \\ \lambda_1 &= 3, & H_1(z) &= ze^{-z^2/2}, \\ \lambda_2 &= 5, & H_2(z) &= (1 - 2z)e^{-z^2/2}, \\ \lambda_3 &= 7, & H_3(z) &= (3z - 2z^3)e^{-z^2/2}. \end{aligned}$$

4.5 INTEGRAL EQUATIONS

For many purposes it is convenient to reformulate boundary value problems in terms of integral equations. We consider the differential equation

$$y'' + [\lambda - q(z)]y = 0, \tag{1}$$

where $q(z)$ is continuous and bounded for $0 \leqslant z \leqslant \pi$, with the separated boundary conditions

$$\begin{aligned} a_0 y(0) + a_1 y'(0) &= 0, \\ b_0 y(\pi) + b_1 y'(\pi) &= 0. \end{aligned} \tag{2}$$

This boundary value problem can be rewritten as a system by letting

$$\begin{aligned} y &= x_1, \\ y' &= x_2, \end{aligned}$$

so that (1) can be rewritten as

$$\begin{pmatrix} x_1' \\ x_2' \end{pmatrix} = X' = AX - \lambda BX, \tag{3}$$

where

$$A = \begin{pmatrix} 0 & 1 \\ q(z) & 0 \end{pmatrix}, \qquad B = \begin{pmatrix} 0 & 0 \\ 1 & 0 \end{pmatrix}.$$

If the vectors

$$V_1 = \begin{pmatrix} a_0 \\ a_1 \end{pmatrix}, \qquad V_2 = \begin{pmatrix} b_0 \\ b_1 \end{pmatrix}$$

are introduced, the boundary conditions (2) are replaced in terms of inner products by

$$\begin{aligned} (V_1, X(0)) &= 0, \\ (V_2, X(\pi)) &= 0. \end{aligned} \tag{4}$$

Next we turn to the equation

$$X' = AX, \tag{5}$$

which is (3) when $\lambda = 0$, and seek two solutions satisfying

$$\begin{aligned} (V_1, X_1(0)) &= 0, \\ (V_2, X_2(\pi)) &= 0. \end{aligned} \tag{6}$$

That such solutions exist follows from the fundamental existence theorems of Chapter 2. We shall also suppose that X_1 and X_2 are linearly independent. This is equivalent to the statement that $\lambda = 0$ is not an eigenvalue. There is no loss in generality in this supposition, since otherwise (1) could be rewritten as

$$y'' + [\lambda' - (q(z) - \mu)]y = 0,$$

where $\lambda' = \lambda - \mu$, and μ could be chosen so that for all eigenvalues, $\lambda' > 0$. One can now define the matrix

$$\Phi = \begin{pmatrix} x_{11} & x_{12} \\ x_{21} & x_{22} \end{pmatrix},$$

where

$$X_1 = \begin{pmatrix} x_{11} \\ x_{21} \end{pmatrix}, \qquad X_2 = \begin{pmatrix} x_{12} \\ x_{22} \end{pmatrix},$$

and introduce new dependent variables into (3) by letting

$$X = \Phi Y.$$

X_1 and X_2 are not uniquely determined, but can be multiplied by arbitrary constants without violating (6). It was seen in Chapter 2, Section 8 that the Wronskian (that is, $|\Phi|$), is given by

$$|\Phi| = c \exp \int \operatorname{tr} A \, dz,$$

but in this case tr $A = 0$, so that $|\Phi|$ is a constant. Therefore the constants can be so selected that $|\Phi| = 1$, unless $|\Phi|$ vanished identically. But in this case X_1 and X_2 would be linearly dependent, which is not so. Then (3) becomes

$$Y' = -\lambda QY, \tag{7}$$

where $Q = \Phi^{-1}B\Phi$.

The boundary conditions (6) now become

$$(V_1, \Phi(0)Y(0)) = (V_1', Y(0)) = 0,$$

where

$$V_1' = \begin{pmatrix} 0 \\ a_0x_{12}(0) + a_1x_{22}(0) \end{pmatrix}, \tag{8}$$

$$(V_2, \Phi(\pi)Y(\pi)) = (V_2', Y(\pi)) = 0,$$

where

$$V_2' = \begin{pmatrix} b_0x_{11}(\pi) + b_1x_{21}(\pi) \\ 0 \end{pmatrix}.$$

An integration of (7) leads to the integral equation

$$Y = c_1\begin{pmatrix} 1 \\ 0 \end{pmatrix} + c_2\begin{pmatrix} 0 \\ 1 \end{pmatrix} - \lambda\int_0^z Q(t)\, Y(t)\, dt. \tag{9}$$

From (8) one can determine the constants of integration, c_1 and c_2. Then

$$c_2 = 0$$

$$c_1 = -\lambda\int_0^\pi x_{12}(x_{11}y_1 + x_{12}y_2)\, dt,$$

where

$$Y = \begin{pmatrix} y_1 \\ y_2 \end{pmatrix}.$$

Finally (9) becomes

$$\begin{aligned} Y = &-\lambda\int_z^\pi \begin{pmatrix} x_{12}(x_{11}y_1 + x_{12}y_2) \\ 0 \end{pmatrix} dt \\ &-\lambda\int_0^z \begin{pmatrix} 0 \\ x_{11}(x_{11}y_1 + x_{12}y_2) \end{pmatrix} dt. \end{aligned} \tag{10}$$

The equation (10) is an integral equation equivalent to (7) with the boundary conditions (8) already built in. One can now return to X and ultimately to the function $y(z)$ in (1). A calculation then shows that

$$y = -\lambda x_{12}(z)\int_0^z x_{11}(t)\, y(t)\, dt - \lambda x_{11}(z)\int_z^\pi x_{12}(t)\, y(t)\, dt.$$

The latter can be written as

$$y = \lambda \int_0^\pi K(z, t)\, y(t)\, dt, \tag{11}$$

where

$$\begin{aligned} K(z, t) &= -x_{12}(z)\, x_{11}(t), \qquad t \leqslant z, \\ &= -x_{12}(t)\, x_{11}(z), \qquad t \geqslant z. \end{aligned} \tag{12}$$

The equation (11) is equivalent to the differential equation (1) with boundary conditions (2). It is an integral equation of a class known as a homogeneous Fredholm integral equation of the second kind. For general values of λ, it has only the solution $y = 0$. But when λ is an eigenvalue, it has nonvanishing solutions. $K(z, t)$ is known as the kernel of the equation.

EXAMPLE 1. The boundary value problem

$$y'' + \lambda y = 0,$$
$$y(0) = y(\pi) = 0,$$

can be rewritten as the system

$$\begin{pmatrix} x_1' \\ x_2' \end{pmatrix} = X' = \begin{pmatrix} 0 & 1 \\ 0 & 0 \end{pmatrix} X - \lambda \begin{pmatrix} 0 & 0 \\ 1 & 0 \end{pmatrix} X$$

with boundary conditions

$$(V_1, X(0)) = 0, \qquad (V_2, X(\pi)) = 0,$$

where

$$V_1 = V_2 = \begin{pmatrix} 1 \\ 0 \end{pmatrix}.$$

In accordance with the general discussion we now use (5) and (6) to find X_1 and X_2:

$$X = c_1 \begin{pmatrix} 1 \\ 0 \end{pmatrix} + c_2 \begin{pmatrix} t \\ 1 \end{pmatrix},$$

so that

$$X_1 = c_3 \begin{pmatrix} t \\ 1 \end{pmatrix},$$

$$X_2 = c_4 \begin{pmatrix} \pi - t \\ -1 \end{pmatrix}.$$

Now c_3 and c_4 must be so selected that the matrix

$$\Phi = \begin{pmatrix} c_3 t & c_4(\pi - t) \\ c_3 & -c_4 \end{pmatrix}$$

has determinant 1. This can be accomplished by letting

$$c_3 = 1, \qquad c_4 = -\frac{1}{\pi}.$$

New dependent variables are introduced by

$$X = \Phi Y.$$

It follows that

$$Y' = -\lambda \begin{pmatrix} \dfrac{(\pi - t)\, t}{\pi} & \dfrac{-(\pi - t)^2}{\pi^2} \\ t^2 & \dfrac{-(\pi - t)\, t}{\pi} \end{pmatrix} Y,$$

and the boundary conditions (8) take the form

$$(V_3, Y(0)) = 0, \qquad (V_4, Y(\pi)) = 0,$$

$$V_3 = \begin{pmatrix} 0 \\ -1 \end{pmatrix}, \qquad V_4 = \begin{pmatrix} \pi \\ 0 \end{pmatrix}.$$

The solution of this system (10) is

$$Y = \lambda \begin{pmatrix} \displaystyle\int_z^\pi \frac{\pi - t}{\pi} \left(t y_1 - \frac{\pi - t}{\pi} y_2 \right) dt \\ \displaystyle -\int_0^z t \left(t y_1 - \frac{\pi - t}{\pi} y_2 \right) dt \end{pmatrix},$$

where

$$Y = \begin{pmatrix} y_1 \\ y_2 \end{pmatrix}.$$

After making suitable transformations to return to the original dependent variable y, one finds

$$\begin{aligned} y &= \lambda \left[\frac{\pi - z}{\pi} \int_0^z t y \, dt + z \int_z^\pi \frac{\pi - t}{\pi} y \, dt \right] \\ &= \lambda \int_0^\pi K(z, t)\, y \, dt, \\ K(z, t) &= \frac{(\pi - z)\, t}{\pi}, \qquad t \leqslant z, \\ &= \frac{(\pi - t)\, z}{\pi}, \qquad t \geqslant z. \end{aligned}$$

The kernel (12) has a very important and interesting interpretation. From (5) it follows that both $x_{11}(z)$ and $x_{12}(z)$ satisfy the differential equation

$$y'' - q(z) y = 0,$$

and $x_{11}(z)$ satisfies the boundary condition at $z = 0$, and $x_{12}(z)$ the one at $z = \pi$. Furthermore $K(z, t)$ is a continuous function of z, but it is not differentiable at

$z = t$. At $z = t$, $(\partial/\partial z)K(z, t)$ has a jump discontinuity. This jump can be calculated as follows: Let

$$\left[\frac{\partial}{\partial z} K(z, t)\right]_{z=t}$$

denote the jump. Then

$$\begin{aligned}\left[\frac{\partial}{\partial z} K(z, t)\right]_{z=t} &= \lim_{z\to t^+} \frac{\partial}{\partial z} K(z, t) - \lim_{z\to t^-} \frac{\partial}{\partial z} K(z, t) \\ &= -x_{12}'(t)\, x_{11}(t) + x_{12}(t)\, x_{11}'(t),\end{aligned}$$

where the limits indicate that $z = t$ be approached through values of z greater and smaller than t, respectively. Clearly $[(\partial/\partial z)K(z, t)]_{z=t}$ is the negative of the wronskian $|\Phi|$, which was so constructed that $|\Phi| = 1$. Hence

$$\left[\frac{\partial}{\partial z} K(z, t)\right]_{z=t} = -1. \tag{13}$$

It follows that the expression

$$\frac{\partial^2}{\partial z^2} K - q(z)\, K$$

must vanish for $z \neq t$ and be unbounded at $z = t$ in such a way that its integral across $z = t$ must have a jump discontinuity and will satisfy (13). A function with such a character often arises in mathematical physics and is called a delta or impulse function and denoted $\delta(z - t)$. The properties of this function can be described as follows:

$$\begin{aligned}\delta(z - t) &= 0, \qquad z \neq t, \\ \int_a^b \delta(z - t)\, dt &= 1, \qquad a < z < b.\end{aligned} \tag{14}$$

The rigorous justification of such functions will not be discussed here, but interested readers may pursue the subject by reading up on the theory of distributions.

It follows that $K(z, t)$ satisfies the inhomogeneous equation

$$\frac{\partial^2}{\partial z^2} K - q(z)\, K = -\delta(z - t), \tag{15}$$

and boundary conditions (2).

An important property of delta functions is symbolized by the following formula:

$$\int_a^b f(t)\, \delta(z - t)\, dt = f(z), \qquad a < z < b, \tag{16}$$

for continuous functions $f(z)$. To prove (16) we write

$$f(t) = f(z) + [f(t) - f(z)].$$

Now

$$\int_a^b f(z)\,\delta(z-t)\,dt = f(z)\int_a^b \delta(z-t)\,dt = f(z)$$

by (14) and

$$\int_a^b [f(t)-f(z)]\,\delta(z-t)\,dt = \int_{z-\varepsilon}^{z+\epsilon} [f(t)-f(z)]\,\delta(z-t)\,dt,$$

since $\delta(z-t)$ vanishes for all t outside all neighborhoods of z. But for continuous $f(z)$ and any number τ it must follow that

$$|f(t)-f(z)| < \tau, \qquad \text{for } |t-z| < \epsilon,$$

provided ϵ is chosen sufficiently small, Then by the mean value theorem (whose applicability here is really justified by the theory of distributions), it follows that

$$\left|\int_{z-\epsilon}^{z+\epsilon} [f(t)-f(z)]\,\delta(z-t)\,dt\right| < \tau\int_{z-\epsilon}^{z+\epsilon} |\delta(z-t)|\,dt = \tau,$$

thus proving (16).

From (15) it is very easy to rederive (12), as follows: Let $x_{11}(z)$ and $x_{12}(z)$ be solutions of (1) with $\lambda=0$, satisfying the boundary conditions at $z=0$ and $z=\pi$, respectively. These are not unique, but can be multiplied with arbitrary scalars. These can then be so selected that the Wronskian of these solutions can be made equal to unity. Therefore

$$x_{11}(z)\,x_{12}'(z) - x_{12}(z)\,x_{11}'(z) = 1.$$

Then we find that

$$\begin{aligned} K(z,t) &= c\,\frac{x_{11}(z)}{x_{11}(t)}, \qquad z \leqslant t, \\ &= c\,\frac{x_{12}(z)}{x_{12}(t)}, \qquad z \geqslant t \end{aligned}$$

satisfies (15) for $z \neq t$ and is evidently continuous at $z=t$. The constant c must now be so selected that $K(z,t)$ satisfies the jump condition (13).

$$\begin{aligned} \left[\frac{\partial}{\partial z} K(z,t)\right]_{z=t} &= c\,\frac{x_{12}'(t)}{x_{12}(t)} - c\,\frac{x_{11}'(t)}{x_{11}(t)} \\ &= \frac{c}{x_{12}(t)\,x_{11}(t)}\,[x_{11}(t)\,x_{12}'(t) - x_{12}(t)\,x_{11}'(t)] \\ &= \frac{c}{x_{12}(t)\,x_{11}(t)} = -1. \end{aligned}$$

Therefore

$$c = -x_{12}(t)x_{11}(t)$$

and

$$\begin{aligned} K(z, t) &= -x_{12}(t)x_{11}(z), \qquad z \leqslant t, \\ &= -x_{12}(z)x_{11}(t), \qquad z \geqslant t, \end{aligned}$$

which was found in (12).

EXAMPLE 2. Consider the problem

$$\frac{\partial^2}{\partial z^2} K(z, t) = -\delta(z - t),$$

$$K(0, t) = K(\pi, t) = 0.$$

We find that for $z < t$ and $z > t$, K has to be a linear function in z. Then, considering the boundary conditions and that K has to be continuous at $z = t$, we find

$$\begin{aligned} K &= c\frac{z}{t}, \qquad z \leqslant t, \\ &= c\frac{\pi - z}{\pi - t}, \qquad z \geqslant t. \end{aligned}$$

To determine c, we use (13):

$$\left[\frac{\partial}{\partial z} K(z, t)\right]_{z=t} = \frac{-c}{\pi - t} - \frac{c}{t} = -1.$$

Therefore

$$c = \frac{t(\pi - t)}{\pi},$$

so that

$$\begin{aligned} K &= \frac{(\pi - t)\, z}{\pi}, \qquad z \leqslant t, \\ &= \frac{(\pi - z)\, t}{\pi}, \qquad z \geqslant t, \end{aligned}$$

which agrees with the result of Example 1.

The function $K(z, t)$ is generally known as the Green's function. A solution of the equation

$$y'' - q(z)y = f(z) \tag{17}$$

with boundary conditions

$$\begin{aligned} a_0 y(0) + a_1 y'(0) &= 0, \\ b_0 y(\pi) + b_1 y'(\pi) &= 0, \end{aligned} \tag{18}$$

can be constructed for arbitrary continuous $f(z)$ by means of the Green's function. We shall now show that the solution of (17) subject to the boundary conditions (18) is given by

$$y = -\int_0^\pi K(z, t)\, f(t)\, dt. \tag{19}$$

To do so, we use (12) to write

$$y = x_{12}(z) \int_0^z x_{11}(t) f(t)\, dt + x_{11}(z) \int_z^\pi x_{12}(t) f(t)\, dt,$$

$$y' = x_{12}'(z) \int_0^z x_{11}(t) f(t)\, dt + x_{11}'(z) \int_z^\pi x_{12}(t) f(t)\, dt,$$

$$y'' = x_{12}''(z) \int_0^z x_{11}(t) f(t)\, dt + x_{11}''(z) \int_z^\pi x_{12}(t) f(t)\, dt$$

$$+ [x_{12}'(z)\, x_{11}(z) - x_{11}'(z)\, x_{12}(z)]\, f(z).$$

Evidently

$$a_0 y(0) + a_1 y'(0) = [a_0 x_{11}(0) + a_1 x_{11}'(0)] \int^\pi x_{12}(t) f(t)\, dt = 0,$$

since x_{11} satisfies the boundary condition at $z = 0$, and similarly one can show that the condition at $z = \pi$ is also satisfied. Since both x_{11} and x_{12} satisfy

$$y'' - q(z) y = 0,$$

it follows that for (19),

$$y'' - q(z)\, y = [x_{11}(z)\, x_{12}'(z) - x_{11}'(z)\, x_{12}(z)]\, f(z) = f(z).$$

If one proceeds in a formal way and differentiates under the integral, one finds by using (19), (15), and (16) that

$$y'' - q(z)\, y = -\int_0^\pi \left[\frac{\partial^2}{\partial z^2} K(z, t) - q(z)\, K(z, t)\right] f(t)\, dt$$

$$= \int_0^\pi \delta(z - t) f(t)\, dt = f(z).$$

If we now return to the example 1 and write it in the form

$$y'' - q(z) y = -\lambda y,$$

it is in the form (17) with

$$f(z) = -\lambda y.$$

Then from (19) the solution of (1) with boundary condition (2) is given by

$$y = \lambda \int_0^\pi K(z, t)\, y(t)\, dt,$$

thereby rederiving (11) by a different method.

It follows that the problem of solving the integral equation (11) and the differential equation (1) with boundary conditions (2) are equivalent. But for many purposes it is more convenient to work with (11) than with the differential equation (1), as will be seen in the next section.

4.6 FUNCTION SPACES

We shall now consider the class S of all real continuous functions defined over the interval $[0, \pi]$, which also satisfy the boundary conditions (2) of the preceding section. Under the standard operations of addition and scalar multiplication, this class forms a linear vector space, as defined in Section 1.8. A verification of these facts is trivial.

Furthermore one can introduce an inner product of two such functions, $f(z)$ and $g(z)$, by

$$(f, g) = \int_0^\pi f(z)\, g(z)\, dz. \tag{1}$$

This definition satisfies the properties (*a*), (*b*), (*c*) of inner products (see Section 3 of Chapter 1). Again a verification of these is immediate. Finally we can define norm $||f||$ by

$$||f|| = (f, f)^{1/2}. \tag{2}$$

A verification of properties (*a*) and (*b*) of norms is immediate (see Section 7 of Chapter 1). Property (*c*) states that

$$||f + g|| \leqslant ||f|| + ||g||. \tag{3}$$

To prove this result, we return to the Schwarz inequality:

$$\left| \sum_{i=1}^n a_i b_i \right|^2 < \sum_{i=1}^n a_i^2 \sum_{i=1}^n b_i^2,$$

which can be generalized from sums to integrals and then becomes

$$\left| \int_0^\pi f(z)\, g(z)\, dz \right|^2 \leqslant \int_0^\pi f^2(z)\, dz \int_0^\pi g^2(z)\, dz.$$

We now observe that

$$\begin{aligned}\int_0^\pi [f(z)+g(z)]^2\,dz &= \int_0^\pi f^2(z)\,dz + 2\int_0^\pi f(z)g(z)\,dz \\ &\quad + \int_0^\pi g^2(z)\,dz \\ &\leqslant \int_0^\pi f^2(z)\,dz + 2\left|\int_0^\pi f(z)g(z)\,dz\right| \\ &\quad + \int_0^\pi g^2(z)\,dz \\ &\leqslant \int_0^\pi f^2(z)\,dz + 2\left[\int_0^\pi f^2(z)\,dz \int_0^\pi g^2(z)\,dz\right]^{1/2} \\ &\quad + \int_0^\pi g^2(z)\,dz \\ &= \left[\left(\int_0^\pi f^2(z)\,dz\right)^{1/2} + \left(\int_0^\pi g^2(z)\,dz\right)^{1/2}\right]^2.\end{aligned}$$

By taking square roots, (3) follows from the last inequality.

We now observe that the transformation

$$\int_0^\pi K(z,t)f(t)\,dt = g(z), \tag{4}$$

denoted symbolically by $Kf = g$, applied to $f(z)$ produces a function $g(z)$. $K(z, t)$ is the Green's function of the preceding section. Since $K(z, t)$ and $f(t)$ are continuous, it follows that $g(z)$ is also a continuous function. Furthermore $K(z, t)$ satisfies the required boundary conditions, and from this it can be shown that $g(z)$ also satisfies the boundary conditions. It follows that the operation (4) is a linear transformation on the linear vector space of continuous functions over the interval $[0, \pi]$. It was seen in (12) of Section 4.5 that $K(z, t)$ is a symmetric function of its arguments; that is,

$$K(z,t) = K(t,z).$$

From this fact it follows that we have here a self-adjoint transformation; that is,

$$(f, Kg) = (Kf, g).$$

To prove this, we use (1) and (4):

$$\begin{aligned}(f, Kg) &= \int_0^\pi f(z)\,dz \int_0^\pi K(z,t)g(t)\,dt \\ &= \int_0^\pi g(t)\,dt \int_0^\pi K(z,t)f(z)\,dz \\ &= \int_0^\pi g(t)\,dt \int_0^\pi K(t,z)f(z)\,dz \\ &= (Kf, g).\end{aligned}$$

From the self-adjointness of the transformation K it follows that all eigenvalues of K must be real; that is, if K has eigenfunctions such that

$$\lambda K y = y,$$

then λ and y must be real. The proof is completely analogous to the comparable case of symmetric matrices and has been also proved in Section 4.4 directly from the differential equation.

We find, therefore, that the equation

$$Ky = \frac{1}{\lambda} y$$

has an infinite set of eigenvalues,

$$\frac{1}{\lambda_0} > \frac{1}{\lambda_1} > \frac{1}{\lambda_2} > \frac{1}{\lambda_3} > \cdots,$$

and eigenfunctions u_i such that

$$Ku_i = \frac{1}{\lambda_i} u_i.$$

Furthermore the eigenfunctions u_i are orthogonal in the sense that

$$(u_i, u_j) = \int_0^\pi u_i(z)\, u_j(z)\, dz = 0, \qquad i = j. \tag{5}$$

Since

$$(u_i, u_i) > 0,$$

the set $\{u_i\}$ can be made orthonormal by selecting the u_i such that

$$(u_i, u_i) = 1. \tag{6}$$

Example 1. The equation

$$Ky = \int_0^\pi K(z, t)\, y(t)\, dt = \frac{1}{\lambda} y(z),$$

where

$$K(z, t) = \frac{(\pi - z)\, t}{\pi}, \qquad t \leqslant z,$$

$$= \frac{(\pi - t)\, z}{\pi}, \qquad t \geqslant z,$$

stems from the boundary value problem

$$y'' + \lambda y = 0,$$
$$y(0) = y(\pi) = 0.$$

It follows that its eigenvalues are given by

$$\lambda_n = n^2$$

and the eigenfunctions by

$$u_n = \left(\frac{2}{\pi}\right)^{1/2} \sin nz.$$

These satisfy (5) and (6).

We have seen that the linear transformation K has a set of eigenvalues and eigenfunctions. In comparable situations in finite dimensional vector spaces, the eigenvectors could be used to construct an orthonormal basis. In the present instance we are dealing with a self-adjoint transformation whose eigenvalues are therefore real and its eigenfunctions are orthonormal. It seems plausible, by analogy, that this orthonormal set forms a basis for the linear vector space S of continuous functions. In that case it would follow that if $f(z)$ is any function in S, then

$$f(z) = \sum_{i=0}^{\infty} c_i u_i(z).$$

To evaluate the c_i, we use the orthonormal character of the basis. Then

$$\begin{aligned}(f(z), u_j(z)) &= \sum_{i=0}^{\infty} c_i(u_i(z), u_j(z)) \\ &= c_j, \qquad j = 0, 1, 2, \cdots, \end{aligned} \tag{7}$$

so that

$$f(z) = \sum_{i=0}^{\infty} (f(z), u_i(z))\, u_i(z). \tag{8}$$

Two important questions arise in connection with the series in (8):

(1) Does the series converge, and in what sense?

(2) If the series does converge in some sense, does it converge to $f(z)$?

We now define the partial sum

$$s_n(z) = \sum_{i=0}^{n} d_i u_i(z) \tag{9}$$

and consider

$$\| f(z) - s_n(z) \| = \left\{ \int_0^{\pi} [f(z) - s_n(z)]^2 \, dz \right\}^{1/2}. \tag{10}$$

We shall now show that among all $s_n(z)$, (10) assumes a minimum value for that $s_n(z)$ for which the coefficients d_i in (9) coincide with the c_i defined in (7). To do so, we let

$$d_i = c_i + e_i.$$

A direct computation shows that

$$\| f(z) - \sum_{i=0}^{n} (c_i + e_i)\, u_i(z) \| = \left\{ \int_0^{\pi} f^2(z)\, dz - \sum_{i=0}^{n} c_i^2 + \sum_{i=0}^{n} e_i^2 \right\}^{1/2}$$

and evidently the right side has a minimum value if all $e_i = 0$, in which case $d_i = c_i$. Since norms are nonnegative, it also follows that

$$\int_0^{\pi} f^2(z)\, dz \geqslant \sum_{i=0}^{n} c_i^2 \tag{11}$$

Inequality (11) is called Bessel's inequality. From it one observes immediately that since the left side is independent of n, the series on the right converges. Furthermore if

$$\int_0^{\pi} f^2(z)\, dz = \sum_{i=0}^{\infty} c_i^2\,, \tag{12}$$

then it follows from (10) that

$$\lim_{n\to\infty} \left\| f(z) - \sum_{i=0}^{n} c_i u_i(z) \right\| = 0. \tag{13}$$

In this case we say that the orthormal expansion converges to $f(z)$ in the mean. The equation (12) is known as Parseval's equation, and is a necessary and sufficent condition for convergence in the mean. The next section will be devoted to a discussion of the convergence of (8)

4.7 FOURIER EXPANSIONS

The series (8) of the preceding section is often called the Fourier series of $f(z)$, and the coefficients c_i are referred to as the Fourier coefficients. To discuss their convergence properties, we must return to the integral equation

$$\int_0^{\pi} K(z, t)\, y(t)\, dt = \frac{1}{\lambda}\, y(z) \tag{1}$$

and discuss some of its properties. We are here concerned with those kernels $K(z, t)$ that arose in connection with Sturm-Liuoville problems, where the function $q(z)$ was continuous in the interval $[0, \pi]$ and the boundary conditions were separated. Under these conditions, $K(z, t)$ was bounded for all z and t in $[0, \pi]$ and symmetric in the arguments z and t.

We now turn to (1) and multiply by $y(z)$ and integrate:

$$\int_0^\pi \int_0^\pi K(z, t)\, y(z)\, y(t)\, dz\, dt = \frac{1}{\lambda} \int_0^\pi y^2(z)\, dz.$$

For $y = u_0(z)$ and $\lambda = \lambda_0$, the above reduces to

$$\int_0^\pi \int_0^\pi K(z, t)\, u_0(z)\, u_0(t)\, dz\, dt = \frac{1}{\lambda_0}.$$

We shall now show that the eigenfunction and eigenvalue can be characterized in the following fashion. The eigenfunction is that function in the set of functions of the linear space S that satisfies

$$\int_0^\pi y^2\, dz = 1 \tag{2}$$

for which the expression

$$\int_0^\pi \int_0^\pi K(z, t)\, y(z)\, y(t)\, dz\, dt \tag{3}$$

attains a maximum value, and $1/\lambda_0$ is given by that maximum value. That (3) has a maximum value follows by an application of the Schwarz inequality:

$$\begin{aligned}
\left| \int_0^\pi y(z)\, dz \int_0^\pi K(z, t)\, y(t)\, dt \right| &\leqslant \int_0^\pi |\, y(z)\,|\, dz \left\{ \int_0^\pi K^2(z, t)\, dt \int_0^\pi y^2(t)\, dt \right\}^{1/2} \\
&= \int_0^\pi |\, y(z)\,|\, dz \left\{ \int_0^\pi K^2(z, t)\, dt \right\}^{1/2} \\
&\leqslant \left\{ \int_0^\pi y^2(z)\, dz \int_0^\pi \int_0^\pi K^2(z, t)\, dz\, dt \right\}^{1/2} \\
&= \left\{ \int_0^\pi \int_0^\pi K^2(z, t)\, dz\, dt \right\}^{1/2}.
\end{aligned}$$

Since $K(z, t)$ is bounded, the right side of the inequality has an upper bound and (3) must have a least upper bound. We shall assume without proof the (nontrivial) fact that there exists a function $u_0(z)$ for which (2) holds, once (3) has attained its maximum value. We now seek to characterize that function $u_0(z)$. The methods used here are borrowed from the calculus of variations.

Consider any function in our class of the form

$$y(z) = u_0(z) + \epsilon\eta(z),$$

where ϵ is a parameter and $\eta(z)$ is a function in our class. Then, from (2), we find

$$\int_0^\pi u_0^2\, dz + 2\epsilon \int_0^\pi u_0\eta\, dz + \epsilon^2 \int_0^\pi \eta^2\, dt = 1,$$

and since u_0 satisfies (2), we have

$$2\epsilon \int_0^\pi u_0\eta \, dz + \epsilon^2 \int_0^\pi \eta^2 \, dz = 0. \tag{4}$$

Since $u_0(z)$ maximizes (3), replacing y by $u_0(z) + \epsilon\eta, (z)$ cannot increase the maximum, so that

$$\int_0^\pi \int_0^\pi K(z, t) \, [u_0(z) + \epsilon\eta(z)] \, [u_0(t) + \epsilon\eta(t)] \, dz \, dt \leqslant \int_0^\pi \int_0^\pi K(z, t) \, u_0(z) \, u_0(t) \, dz \, dt.$$

From the preceding equation we obtain

$$2\epsilon \int_0^\pi \eta(z) \, dz \int_0^\pi K(z, t) \, u_0(t) \, dt + \epsilon^2 \int_0^\pi \int_0^\pi K(z, t) \, \eta(z) \, \eta(t) \, dz \, dt \leqslant 0. \tag{5}$$

Since (4) and (5) must hold for arbitrary ϵ, and the linear terms will govern the behavior of these functions, it follows that we require that

$$\int_0^\pi \eta(z) \, u_0(z) \, dz = 0, \tag{6}$$

$$\int_0^\pi \eta(z) \, dz \int_0^\pi K(z, t) \, u_0(t) \, dt = 0. \tag{7}$$

The set of all functions $u_0(z)$ that can satisfy (6) for arbitrary $\eta(z)$, none of which are multiples of $u_0(z)$, must contain only one linearly independent element. Otherwise, suppose we had in addition to (6),

$$\int_0^\pi \eta(z) \, v(z) \, dz = 0. \tag{8}$$

Then for $\eta(z) = v(z)$, (6) would hold and (8) would be violated; and for $\eta(z) = u_0(z)$, (6) would be violated. Therefore any two functions satisfying (6) must be linearly dependent, and we conclude from (6) and (7) that

$$\int_0^\pi K(z, t) \, u_0(t) \, dt = \frac{1}{\lambda_0} u_0(z),$$

where the constant must be chosen so as to agree with the eigenvalue corresponding to $u_0(z)$. In this case (3) becomes

$$\int_0^\pi \int_0^\pi K(z, t) \, u_0(z) \, u_0(t) \, dz \, dt = \frac{1}{\lambda_0} \int_0^\pi u_0^2(t) \, dt = \frac{1}{\lambda_0},$$

and since λ_0 is the smallest eigenvalue, (3) has attained its maximum value.

The above argument, incidentally, shows that if $K(z, t)$ is any symmetric function for which (3) attains a maximum value, then the equation

$$\int_0^\pi K(z, t)\, y(t)\, dt = \frac{1}{\lambda} y(z)$$

has a maximum eigenvalue $1/\lambda_0$.

One can characterize by similar arguments the next eigenvalue and eigenfunctions by maximizing (3) subject to (2) and imposing the additional constraint

$$\int_0^\pi y(z)\, u_0(z)\, dz = 0.$$

In general the n'th eigenvalue and eigenfunction are found by maximizing (3) subject to (2) and imposing the constraints

$$\int_0^\pi y(z)\, u_i(z)\, dz = 0, \qquad i = 0, 1, 2, \cdots, n-1.$$

If we proceed formally and seek to expand $K(z, t)$ as a function of z in a Fourier series, we find that

$$K(z, t) = \sum_{i=0}^{\infty} c_i u_i(z),$$

where

$$c_i = \int_0^\pi K(z, t)\, u_i(z)\, dz = \frac{1}{\lambda_i} u_i(t).$$

The resultant series,

$$\sum_{i=0}^{\infty} \frac{u_i(z)\, u_i(t)}{\lambda_i}, \tag{9}$$

converges absolutely by virtue of the asymptotic formulas (8) and (28) of Section 4.4. This follows from a comparison to the series

$$\sum_1^\infty \frac{1}{n^2}.$$

But we must show that (9) does indeed converge to $K(z, t)$. We now consider the new kernel

$$R(z, t) = K(z, t) - \sum_{i=0}^{\infty} \frac{u_i(z)\, u_i(t)}{\lambda_i} \tag{10}$$

and the integral equation

$$\mu \int_0^\pi R(z, t)\, y(t)\, dt = y(z). \tag{11}$$

From the preceding discussion it follows that, since

$$R(z, t) = R(t, z), \qquad \int_0^\pi \int_0^\pi R^2(z, t)\, dz\, dt < \infty,$$

(11) has a smallest eigenvalue and a corresponding eigenfunction. We shall now state and prove the following lemma.

LEMMA

(1) $\int_0^\pi R(z, t)\, u_i(t)\, dt = 0, \qquad i = 0, 1, 2, \cdots.$

(2) If $y(z)$ satisfies (11), then

$$\int_0^\pi y(z)\, u_i(z)\, dz = 0, \qquad i = 0, 1, 2, \cdots.$$

(3) If y is an eigenfunction of (11) it is also an eigenfunction of (1).

To prove part 1 of the lemma we observe that, using (10),

$$\begin{aligned}\int_0^\pi R(z, t)\, u_i(t)\, dt &= \int_0^\pi K(z, t)\, u_i(t)\, dt \\ &\quad - \sum_{j=0}^{\infty} \frac{u_j(z)}{\lambda_j} \int_0^\pi u_j(t)\, u_i(t)\, dt \\ &= \frac{1}{\lambda_i} u_i(z) - \frac{1}{\lambda_i} u_i(z) = 0.\end{aligned}$$

To prove part 2 of the lemma, we use (11) and part 1 of the lemma and find that

$$\begin{aligned}\int_0^\pi y(z)\, u_i(z)\, dz &= \int_0^\pi u_i(z)\, dz\, \mu \int_0^\pi R(z, t)\, y(t)\, dt \\ &= \mu \int_0^\pi y(t)\, dt \int_0^\pi R(z, t)\, u_i(z)\, dz = 0.\end{aligned}$$

For the proof of part 3 we need part 2. Then

$$\begin{aligned}y(z) &= \mu \int_0^\pi R(z, t)\, y(t)\, dt \\ &= \mu \int_0^\pi K(z, t)\, y(t)\, dt \\ &\quad -\mu \sum_{i=0}^{\infty} \frac{u_i(z)}{\lambda_i} \int_0^\pi u_i(t)\, y(t)\, dt \\ &= \mu \int_0^\pi K(z, t)\, y(t)\, dt.\end{aligned}$$

From part 3 of the lemma we see that y is an eigenfunction of (1), but part 2 shows that it is orthogonal to all eigenfunctions of (1), which is impossible unless y vanishes identically. Therefore (11) has no eigenfunctions. But since an equation of type (11) must have at least one eigenvalue and one eigenfunction, we are led to the conclusion that R must vanish identically. In this case we have

$$K(z, t) = \sum_{i=0}^{\infty} \frac{u_i(z)\, u_i(t)}{\lambda_i}. \tag{12}$$

Thus it has been shown that the kernel $K(z, t)$ can be expanded in a Fourier series, which converges absolutely to $K(z, t)$.

We are now in a position to state and prove the Hilbert-Schmidt theorem. It will not be stated in its fullest generality, but this formulation will be adequate for our immediate purposes.

HILBERT-SCHMIDT THEOREM. If $f(z)$ can be written in the form

$$f(z) = \int_0^{\pi} K(z, t)\, g(t)\, dt, \tag{13}$$

where $g(t)$ is a continuous function in the space S over $[0, \pi]$, and $K(z, t)$ is a symmetric kernel that arises in a Sturm-Liouville problem with a continuous $q(z)$, then $f(z)$ has a Fourier expansion

$$f(z) = \sum_{i=0}^{\infty} c_i u_i(z). \tag{14}$$

This expansion converges to $f(z)$ for all z in $[0, \pi]$.

From (13) we find that

$$\begin{aligned} c_i &= \int_0^{\pi} f(z)\, u_i(z)\, dz = \int_0^{\pi} u_i(z)\, dz \int_0^{\pi} K(z, t)\, g(t)\, dt \\ &= \int_0^{\pi} g(t)\, dt \int_0^{\pi} K(z, t)\, u_i(z)\, dz = \int_0^{\pi} \frac{g(t)\, u_i(t)}{\lambda_i}\, dt. \end{aligned}$$

Then

$$\begin{aligned} \left| f(z) - \sum_{i=0}^{n} c_i u_i(z) \right|^2 &= \left| \int_0^{\pi} \left(K(z, t) - \sum_{i=0}^{n} \frac{u_i(t)\, u_i(z)}{\lambda_i} \right) g(t)\, dt \right|^2 \\ &\leqslant \int_0^{\pi} \left[K(z, t) - \sum_{i=0}^{n} \frac{u_i(t)\, u_i(z)}{\lambda_i} \right]^2 dt \int_0^{\pi} g^2(t)\, dt \end{aligned}$$

by the Schwarz inequality. From the convergence of (12) it follows that the right side of the above inequality must become arbitrarily small as n grows to infinity. Therefore

$$\lim_{n\to\infty} \sum_{i=0}^{n} c_i u_i(z) = f(z).$$

We use this theorem to prove that any function with a continuous second derivative can be expanded in a convergent Fourier series. Suppose $f(z)$ has a continuous second derivative. Then

$$f''(z) - q(z)f(z) = g(z)$$

is a continuous function. The solution of this equation is given by (13) and (14) must hold. By a double integration by parts, it follows that

$$c_n = \int_0^{\pi} f(z)\, u_n(z)\, dz = \int_0^{\pi} f''(z) \left(\int\int u_n(z)\, dz\, dz\right) dz.$$

For large n,

$$u_n(z) \approx \sin nz,$$

so that

$$\int\int u_n(z)\, dz\, dz \approx \frac{-\sin nz}{n^2}.$$

From this it follows that $c_n = 0(1/n^2)$, and it can then be shown that the Fourier series for $f(z)$ converges uniformly.

In the case of the functions that are merely continuous but not necessarily differentiable, one proceeds in a slightly different way. We shall use, without proof, the Weierstrass approximation theorem, which states that any continuous function over the interval $[0, \pi]$ can be approximated by polynomials to any desired degree of accuracy. That is, given $f(z)$ and an arbitrary positive number ϵ, one can find a polynomial $p(z)$ such that

$$|f(z) - p(z)| < \epsilon, \qquad \text{for } 0 \leqslant z \leqslant \pi.$$

We now define

$$s_n(z) = \sum_{i=0}^{n} c_i u_i(z),$$

where

$$c_i = \int_0^{\pi} f(z)\, u_i(z)\, dz;$$

and

$$\sigma_n(z) = \sum_{i=0}^{n} d_i u_i(z),$$

where

$$d_i = \int_0^{\pi} u_i(z)\, p(z)\, dz.$$

Clearly we have

$$\int_0^\pi (f(z) - p(z))^2\, dz < \pi\epsilon^2$$

$$\int_0^\pi (s_n(z) - \sigma_n(z))^2\, dz = \int_0^\pi \left[\sum_{i=0}^n (c_i - d_i)\, u_i(z)\right]^2 dz$$

$$= \sum_{i=0}^n (c_i - d_i)^2 \leqslant \int_0^\pi [f(z) - p(z)]^2\, dz \leqslant \pi\epsilon^2$$

by Bessel's inequality, and

$$\int_0^\pi [p(z) - \sigma_n(z)]^2\, dz \leqslant \pi\epsilon^2,$$

provided n is sufficiently large. The latter follows because $p(z)$ has a continuous second derivative and therefore a uniformly convergent Fourier expansion. It follows now that

$$\begin{aligned} \| f(z) - s_n(z) \| &= \| f(z) - p(z) + p(z) - \sigma_n(z) + \sigma_n(z) - s_n(z) \| \\ &\leqslant \| f(z) - p(z) \| + \| p(z) - \sigma_n(z) \| \\ &\quad + \| \sigma_n(z) - s_n(z) \| < 3\pi^{1/2}\, \epsilon, \end{aligned}$$

so that

$$\lim_{n\to\infty} \| f(z) - s_n(z) \| \leqslant 3\pi^{1/2}\epsilon.$$

Since ϵ is arbitrary, it must follow that

$$\lim_{n\to\infty} \| f(z) - s_n(z) \| = 0.$$

Therefore, if $f(z)$ us any continuous function, then its Fourier series converges to $f(z)$ in the mean.

All these results can be extended to arbitrary symmetric kernels, which satisfy the square integrability condition

$$\int_0^\pi \int_0^\pi K^2(z, t)\, dz\, dt < \infty.$$

In this case any $f(z)$ such that

$$\int_0^\pi f^2(z)\, dz < \infty$$

has a Fourier series that converges to it in the mean.

These expansion theorems are useful in the solution of inhomogeneous boundary value problems of the type

$$\begin{aligned} y'' + [\lambda - q(z)]\,y &= f(z), \\ a_0 y(0) + a_1 y'(0) &= 0, \\ b_0 y(\pi) + b_1 y'(\pi) &= 0. \end{aligned} \tag{15}$$

This problem can be reformulated by solving the equation

$$y'' - q(z)y = -\lambda y + f(z).$$

Then

$$y(z) = \lambda \int_0^\pi K(z, t)\,y(t)\,dt - \int_0^\pi K(z, t) f(t)\,dt \tag{16}$$

is an inhomogeneous Fredholm integral equation. By letting

$$f(z) = \sum_{i=0}^{\infty} f_i u_i(z),$$

$$y = \sum_{i=0}^{\infty} c_i u_i(t),$$

either in (15) or, by a similar process, in (16), one finds

$$\begin{aligned} y'' + [\lambda - q(z)]\,y &= \sum_{i=0}^{\infty} c_i\,[u_i'' - q(z)\,u_i + \lambda u_i] \\ &= \sum_{i=0}^{\infty} c_i\,[\lambda - \lambda_i]\,u_i = \sum_{i=0}^{\infty} f_i u_i, \end{aligned}$$

so that

$$c_i = \frac{f_i}{\lambda - \lambda_i}.$$

It follows that if λ is not an eigenvalue, the solution of (15) and (16) is uniquely given by

$$y = \sum_{i=0}^{\infty} \frac{u_i(z)}{\lambda - \lambda_i} \int_0^\pi f(z)\,u_i(z)\,dz. \tag{17}$$

When λ is an eigenvalue (say, λ_k), there will be either no solution or an infinity of solutions. The latter occurs if and only if $f_k = 0$. Then

$$y = \sum_{\substack{i=0 \\ i \neq k}}^{\infty} \frac{u_i(z)}{\lambda - \lambda_i} \int_0^\pi f(z)\,u_i(z)\,dz + \alpha u_k(z), \tag{18}$$

where α is arbitrary. This result is usually called the Fredholm alternative and deserves to be stated as a theorem.

ALTERNATIVE THEOREM The inhomogeneous integral equation (16), or equivalently the boundary value problem (15), has a unique solution when the corresponding homogeneous problem has no nonvanishing solution. When the homogeneous equation has a nonvanishing solution, (16) has either no solution or an infinity of solutions, depending on the nature of $f(z)$.

The preceding theorem should be compared to the corresponding result in the theory of linear algebraic equations. There one finds that an inhomogeneous system of n linear equations has a unique solution if the corresponding homogeneous system has no nonvanishing solution. This is the case when the determinant of the system does not vanish. When the determinant does vanish, the system has either no solution or an infinity, provided the inhomogeneous terms satisfy a suitable compatibility condition.

4.7 GENERALIZATIONS

In this chapter only the simplest boundary value problem has been treated. The differential equation is of the second order, its coefficients are regular functions, and the boundary points are finite points. Furthermore only the self-adjoint case has been considered. There is a complete but quite complicated theory for self-adjoint problems involving singular coefficients and infinite intervals. In the nonself-adjoint case there is as yet no complete theory in existence.

To obtain even greater generality, one should deal with n'th order boundary value problems. To a very large degree the methods and ideas discussed in this chapter apply to these problems. Whenever such a problem can be reformulated in terms of an integral equation with a kernel $K(z, t)$ such that

$$\int_0^\pi \int_0^\pi K^2(z, t)\, dz\, dt < \infty$$

$$K(z, t) = K(t, z),$$

the preceding results hold with some modifications.

The following problems indicate some of the many possibilities that can arise. Consider the equation

$$y^{(\mathrm{IV})} + \lambda y'' = 0,$$

whose general solution is given by

$$y = c_1 + c_2 z + c_3 \sin \sqrt{\lambda}\, z + c_4 \cos \sqrt{\lambda}\, z.$$

For the boundary conditions,

$$y(0) = y''(0) = y(\pi) = y'(\pi) = 0,$$

we find that we obtain solutions if λ is a root of the equation

$$\tan \sqrt{\lambda}\, \pi = \sqrt{\lambda}\, \pi.$$

This equation has an infinity of real roots $\lambda_1, \lambda_2, \cdots$, and the resultant solutions are

$$y_n = \sin \sqrt{\lambda_n}\, z - z \frac{\sin \sqrt{\lambda_n}\, \pi}{\pi}.$$

For the boundary conditions

$$y''(0) = y'''(0) = y(\pi) = y'(\pi) = 0,$$

we find that there are no eigenvalues, and the only solution is

$$y = 0.$$

For the boundary conditions

$$y(0) = y'(0) = y''(0) = y(\pi) = 0,$$

we find that λ must be a root of

$$\sin \sqrt{\lambda}\, \pi = \sqrt{\lambda}\, \pi.$$

But the solutions of this equation are complex.

Problems

1. Reduce the following equations to the Liouville normal form:

(a) $(1 - z^2)y'' - 2zy' + \lambda y = 0.$

(b) $y'' + (e^{2z} + \lambda)y = 0.$

(c) $y'' + \lambda z^{2a-2} y = 0.$

(d) $z^2 y'' + (2\alpha - 2\nu\beta + 1)zy' + [\beta^2 \lambda z^{2\beta} + \alpha(\alpha - 2\alpha\beta)]y = 0.$

2. Show that the equation

$$y'' + [\lambda + q(z)]y = 0$$

under the substitution

$$\zeta = \int_0^z \frac{dt}{M^2(t)}, \qquad u(\zeta) = \frac{y(z)}{M(z)};$$

where $M(z)$ is a positive solution of

$$M'' + q(z)M = 0,$$

reduces to

$$\frac{d^2u}{d\zeta^2} + \lambda M^4(z(\zeta))\, u = 0.$$

3. With the differential operator

$$Ly = y'' + p(z)y' + q(z)y,$$

one can associate an adjoint differential operator

$$Mu = u'' - (p(z)u)' + q(z)u.$$

In order for the operator M to qualify as the operator adjoint to L, it is necessary that $uLy - yMu$ be an exact derivative; that is,

$$uLy - yMu = \frac{d}{dz}[uy' - yu' + p(z)\, uy].$$

If L and M are identical, we say that L is in self-adjoint form. Show that the operator

$$Ly = y'' + [\lambda - q(z)]y$$

is in self-adjoint form.

4. Show that the operator adjoint to $Ly = y^{(n)} + p_1(z)y^{(n-1)} + p_2(z)y^{(n-2)} + \cdots + p_n(z)y$ is given by

$$Mu = (-1)^n u^{(n)} + (-)^{(n-1)}(p_1(z)u)^{(n-1)} + (-1)^{n-2}(p_2(z)u)^{(n-2)} + \cdots + p_n(z)u.$$

5. Discuss the eigenvalues and eigenfunctions of the boundary value problem

$$y'' + \frac{\lambda}{(z+1)^2}y = 0,$$

$$a_0y(0) + a_1y'(0) = 0,$$

$$b_0y(\pi) + b_1y'(\pi) = 0.$$

6. In Section 4.4 the equation

$$y'' + [\lambda - q(z)]y = 0$$

was discussed and it was assumed that $q(z)$ was continuous and bounded for $0 \leqslant z \leqslant \pi$. Show that the conclusions remain valid, if all that is assumed is that $q(z)$ is integrable; that is,

$$\int_0^\pi |q(z)|\,dz$$

is finite.

7. Discuss the eigenvalues of the equation

$$y'' + [\lambda q(z)]y = 0,$$
$$y(0) = y(\pi) = 0,$$

where

$$\begin{aligned} q(z) &= a > 0, \qquad 0 \leqslant z \leqslant l \\ &= b > 0, \qquad l < z \leqslant \pi. \end{aligned}$$

8. Use the type of argument used in Section 4.3 to show that if z_1 and z_2 are two consecutive zeros of y, which is a solution of

$$\frac{d}{dz} p_1(z)\, y' + q_1(z)\, y = 0,$$

where $p_1(z)$ and $q_1(z)$ are positive functions, then every solution of

$$\frac{d}{dz} p_2(z)\, y' + q_2(z)\, y = 0$$

has to vanish at least once in the interval (z_1, z_2), provided $p_1 \geqslant p_2 > 0$ and $0 < q_1 \leqslant q_2$.

9. Suppose $\lambda(L)$ is a particular eigenvalue of the problem

$$y'' + [\lambda - q(z)]y = 0,$$
$$y(0) = y(L) = 0.$$

Show that as L increases, $\lambda(L)$ must decrease.

10. Consider the wave equation in spherical coordinates:

$$\frac{1}{r^2}\frac{\partial}{\partial r} r^2 \frac{\partial S}{\partial r} + \frac{1}{r^2 \sin^2 \phi}\frac{\partial^2 S}{\partial \theta^2} + \frac{1}{r^2 \sin\phi}\frac{\partial}{\partial \phi} \sin\phi \frac{\partial S}{\partial \phi} = \frac{1}{c^2}\frac{\partial^2 S}{\partial t^2}.$$

Apply the method of separation of variables; that is, let $S(r, \theta, \phi, t) = R(r)\Theta(\theta)\Phi(\phi)T(t)$, and obtain four ordinary differential equations in the independent variables r, θ, ϕ, t. Reduce these to Liouville normal form.

11. Check carefully all the steps leading from (1) to (12) in Section 4.5, filling in all details.

12. Show that the equation

$$y'' - q(z)y = 0,$$

where $q(z) > 0$, and $y(0) > 0$, $y'(0) > 0$, has solutions which are increasing for $z > 0$. Show that if $y(0) > 0$, $y'(0) = 0$, the solution can never vanish for $-\infty < z < \infty$.

13. Show that every solution of

$$y'' + q(z)y = 0,$$

where

$$\int_a^\infty q(z)\, dz = \infty,$$

has an infinite number of zeros.

14. Show that the smallest eigenvalue of the problem

$$y'' + \lambda q(z)y = 0,$$
$$y(0) = y(\pi) = 0,$$

where $q(z) \geqslant 0$ and integrable, must be greater than unity if

$$\pi \int_0^\pi q(z)\, dz \leqslant 4.$$

Hint: First show that

$$4(y^2)_{\max} = \left(\int_0^\pi |y'|\, dz\right)^2.$$

15. Consider the equation

$$(py')' + (\lambda r - q)y = 0$$

with the boundary conditions

$$a_1 p(0)y'(0) + a_0 y(0) = 0,$$
$$b_1 p(\pi)y'(\pi) + b_0 y(\pi) = 0,$$

where p, q, r are continuous and bounded functions, and p and r are positive. Show that the above represents a self-adjoint problem.

(a) Using the notation of Section 4.5, let x_{11} be a solution of

$$(py')' - qy = 0$$

satisfying the boundary condition at $z = 0$, and let x_{12} be another satisfying the boundary condition at $z = \pi$. Assume that these are linearly independent. Show that one can adjust the multiplicative constants so that the Wronskian is given by

$$x_{11}x_{12}' - x_{12}x_{11}' = \frac{1}{p}.$$

(b) Show that the corresponding Green's function is given by

$$\begin{aligned} K(z, t) &= -x_{12}(t)x_{11}(z), \qquad z \leqslant t, \\ &= -x_{12}(z)x_{11}(t), \qquad z \geqslant t. \end{aligned}$$

(c) Show that the above eigenvalue problem can be formulated as an integral equation

$$y = \lambda \int_0^\pi K(z, t)\, r(t)\, y(t)\, dt.$$

(d) The kernel in the above integral is no longer symmetric as was demanded in Section 4.6, but by introducing the new variable

$$x(z) = \sqrt{r(z)}\, y(z)$$

and the new kernel

$$\kappa(z, t) = \sqrt{r(z)}\, K(z, t)\, \sqrt{r(t)},$$

show that the above reduces to an integral equation with symmetric kernel.

(e) Apply the results of the foregoing problem to reduce the equation

$$\begin{aligned} y'' + \lambda z y &= 0, \\ y(0) = y(\pi) &= 0, \end{aligned}$$

to formulate it as an integral equation. Also reduce the above to the Liouville normal form.

16. Show that the eigenfunctions of the boundary value problem in problem 15, y_n, are complete in the space of real and continuous functions under the inner product and norm

$$\begin{aligned} (f, g) &= \int_0^\pi r(z) f(z)\, g(z)\, dt, \\ \| f \| &= (f, f)^{1/2}. \end{aligned}$$

17. In Section 4.7 eigenvalues are characterized in terms of a certain maximum property. For example, for the problem

$$\begin{aligned} y'' + \lambda y &= 0, \\ y(0) = y(1) &= 0, \end{aligned}$$

one can see that the smallest eigenvalue is given by $\lambda_0 = \pi^2$. The associated integral equation is

$$y = \lambda \int_0^1 K(z, t)\, y(t)\, dt,$$

$$\begin{aligned} K(z, t) &= z(1 - t), \qquad z \leqslant t, \\ &= t(1 - z), \qquad z \geqslant t. \end{aligned}$$

According to Section 4.7 (see (2), (3)),

$$\frac{1}{\lambda_0} \geqslant \frac{\int_0^1 \int_0^1 K(z, t)\, y(z)\, y(t)\, dz\, dt}{\int_0^1 y^2(z)\, dz}.$$

Equality in the above can be attained only for the eigenfunction $y = \sin \pi z$. If we let $y = 1$ on the right, we find that

$$\frac{1}{\pi^2} \geqslant \frac{1}{12},$$

so that

$$\pi \leqslant \sqrt{12} = 3.46.$$

The above inequality can be made the basis for a numerical procedure for approximate evaluations of eigenvalues. This is known as the Rayleigh-Ritz method.

Obtain a closer estimate of π by using $y = z(1 - z)$, which function also satisfies the boundary conditions of the problem.

18. Use the preceding problem to find an approximate eigenvalue for the boundary value problem

$$y'' + \lambda z y = 0,$$
$$y(0) = y(\pi) = 0.$$

The above can be solved in terms of Bessel functions. Use tabulated data to check your answer.

19. Consider the integral equation

$$y = \lambda \int_0^1 K(z, t)\, y\, dt,$$

where $K(z, t) = \phi(z)\phi(t)$. Rewriting it in the form

$$y = \phi(z) \left\{ \lambda \int_0^1 \phi(t)\, y(t)\, dt \right\}$$

clearly shows that the solution must be of the form

$$y = m\phi(z)$$

for a suitable constant m. Insertion of this in the integral equation leads to the equation

$$m\left\{1 - \lambda \int_0^1 \phi^2(t)\, dt\right\} = 0.$$

We see that m must vanish unless

$$\lambda = \left\{\int_0^1 \phi^2(t)\, dt\right\}^{-1}.$$

We see that this equation has precisely one eigenvalue.

The inhomogeneous equation

$$y - \lambda \int_0^1 K(z, t)\, y\, dt = f(z)$$

has the solution

$$y = f(z) + m\phi(z),$$

$$m = \frac{\lambda \int_0^1 f(t)\,\phi(t)\, dt}{1 - \lambda \int_0^1 \phi^2(t)\, dt}.$$

This solution exists and is unique unless λ is equal to the eigenvalue

$$\left\{\int_0^1 \phi^2(t)\, dt\right\}^{-1}.$$

In the latter case no solution can exist unless the compatibility condition

$$\int_0^1 f(t)\,\phi\, dt)\, dt = 0$$

is met. If it is, m is arbitrary and an infinity of solutions can be found.

(a) Verify all the above statements in detail.

(b) Generalize the above discussion to the equation

$$y - \lambda \int_0^1 K(z, t)\, y\, dt = f(z),$$

where

$$K(z, t) = \sum_{i=1}^{n} \phi_i(z)\,\phi_i(t).$$

In this case n eigenvalues will in general be found.

(c) Construct some simple equations of the above type and solve them. For example,

$$y - \lambda \int_0^1 (zt + \sin \pi z \sin \pi t)\, y\, dt = 1.$$

20. According to the general theory of Fourier series, a function of period 2π can be expanded into a series of the form

$$f(z) = \frac{a_0}{2} + \sum_1^\infty (a_n \cos nz + b_n \sin nz).$$

This theory does not fit under the framework of Sections 4.6 and 4.7, since there, only boundary value problems with separated boundary conditions were treated. But to obtain the above, we require periodicity boundary conditions. Show how the above could be derived from problems with separated boundary conditions by first writing

$$f(z) = \frac{f(z) + f(-z)}{2} + \frac{f(z) - f(-z)}{2}.$$

The above splits $f(z)$ into the sum of an even and an odd function.

21. Consider the equation

$$y'' + p(z)y' + a^2 y = 0,$$

where a is a constant and $p(z)$ is a positive function such that

$$\int_0^\infty p(z)\, dz = \infty.$$

Show that for all solutions of the equation we have

$$\lim_{z \to \infty} y = 0.$$

Hint. Use (17) of Section 4.4.

5 Linear Differential Equations with Periodic Coefficients

5.1 LINEAR SYSTEMS WITH PERIODIC COEFFICIENTS

The systems to be discussed in this section arise in many practical problems. As will be seen later, many questions regarding nonlinear differential equations with periodic coefficients can be resolved by use of linear systems. Such problems arise, for example, in the study of periodic orbital motion of planets and satellites.

We shall first examine the relatively simple first order equation

$$y' = a(z)y, \tag{1}$$

where $a(z)$ has period T; that is,

$$a(z + T) = a(z).$$

If the mean value of $a(z)$ is given by

$$\frac{1}{T}\int_0^T a(t)\,dt = c,$$

then we can write

$$a(z) = c + p(z),$$

where $p(z)$ has period T and mean value zero. By a change of variable it follows that

$$\int_z^{z+T} p(t)\,dt = 0$$

for all z. The solution of (1) can be written down by inspection so that

$$y(z) = y_0 \exp\left(\int_0^z a(t)\,dt\right) = y_0 \exp\left(cz + \int_0^z p(t)\,dt\right)$$

Clearly,

$$\begin{aligned} y(z + T) &= y_0 \exp\left(cz + cT + \int_0^{z+T} p(t)\,dt\right) \\ &= e^{cT} y_0 \exp\left(cz + \int_0^z p(t)\,dt + \int_z^{z+T} p(t)\,dt\right) \\ &= \mu y_0 \exp\left(cz + \int_0^z p(t)\,dt\right) = \mu y(z) \end{aligned} \tag{2}$$

where

$$\mu = e^{cT}. \tag{3}$$

It follows that the solution of (1) will in general not have period T unless $\mu = 1$. In other words, the necessary and sufficient condition for all solutions of (1) having period T is that $c = 0$, or equivalently, that $a(z)$ have mean value zero.

There is another way of arriving at (2). This method is more general, since the explicit solution of (1) is not required, but it does not lead to the explicit form of μ given in (3). We observe that from (1),

$$y'(z + T) = a(z + T)y(z + T) = a(z)y(z + T),$$

and it follows that $y(z + T)$ is a solution of (1) with initial value $y(T)$. Then

$$y(z + T) = y(z)\frac{y(T)}{y(0)}. \tag{4}$$

Evidently, from the preceding results,

$$\frac{y(T)}{y(0)} = e^{cT} = \mu.$$

The latter procedure has direct applicability to systems. We now consider

$$X' = A(z)X, \tag{5}$$

where $A(z)$ is an $n \times n$ matrix of period T. Suppose $\Phi(z)$ is a fundamental matrix solution of (5); that is, a nonsingular matrix each of whose columns is a solution of (5) and such that

$$\Phi(0) = I.$$

In that case

$$\Phi'(z) = A(z)\Phi(z), \tag{6}$$

and as before,

$$\Phi'(z + T) = A(z + T)\Phi(z + T) = A(z)\Phi(z + T).$$

Now, since $\Phi(z + T)$ is again such a fundamental matrix with initial value $\Phi(T)$, it follows that

$$\Phi(z + T) = \Phi(z)\Phi(T),$$

since the expression on the right satisfies (6) and has initial value $\Phi(T)$. This result is analogous to (4) for the case of a first-order equation.

We shall now assume that $\Phi(T)$ has distinct eigenvalues. Then the equation

$$|\Phi(T) - \mu I| = 0$$

has distinct roots $\mu_1, \mu_2, \cdots \mu_n$. In this case there must exist a nonsingular matrix S such that

$$S^{-1}\Phi(T)S = D$$

is diagonal. We can now define a new fundamental matrix solution (6) by

$$\psi(z) = \Phi(z)S,$$

for which

$$\psi(z + T) = \Phi(z + T)S = \Phi(z)\Phi(T)S = \Phi(z)SD = \psi(z)D. \tag{7}$$

If $X_1, X_2, \cdots X_n$ denote the vector solutions of (5), and if these are also the vector columns of $\psi(z)$, then evidently

$$X_i(z + T) = \mu_i X_i(z), \tag{8}$$

which generalizes (2) to systems. The numbers c_i will be defined by

$$\mu_i = e^{c_i T}.$$

Then, if the matrix $R(z)$ is a diagonal matrix with diagonal elements $e^{c_i z}$, it follows that

$$P(z) = \psi(z)R^{-1}(z) \tag{9}$$

has period T.

$$P(z + T) = \psi(z + T)R^{-1}(z + T) = \psi(z)DR^{-1}(T)R^{-1}(z) = P(z),$$

since

$$DR^{-1}(T) = I.$$

The explicit computation of the μ_i and c_i is in general not possible. But some information regarding the product of all μ_i can be obtained:

$$\mu_1\mu_2 \cdots \mu_n = |D| = |\Phi(T)| = \exp\left(\int_0^T \operatorname{tr} A(t)\, dt\right),$$

and therefore

$$\sum_{i=1}^{n} c_i = \frac{1}{T}\int_0^T tr\, A(t)\, dt. \tag{10}$$

It follows that whenever $\Phi(T)$ can be diagonalized, one obtains n linearly independent solutions, which satisfy (8). In other words, these have the form

$$X_i(z) = e^{c_i z} P_i(z), \tag{11}$$

where the $P_i(z)$ are functions of period T. The case where $\Phi(T)$ cannot be diagonalized can be handled in a similar manner except that $\Phi(T)$ has to be reduced to the Jordan canonical form. Then one can find a matrix S such that

$$S^{-1}\Phi(T)S = J,$$

and as in (7), the matrix

$$\psi(z) = \Phi(z)S$$

satisfies

$$\psi(z + T) = \psi(z)J. \tag{12}$$

If μ is an eigenvalue of multiplicity k, but only one solution of type (11) can be found, then from the structure of J it follows that one can find k solutions such that

$$\begin{aligned} X_1(z + T) &= \mu X_1(z), \\ X_2(z + T) &= \mu X_2(z) + X_1(z), \\ &\vdots \\ X_k(z + T) &= \mu X_k(z) + X_{k-1}(z). \end{aligned} \tag{13}$$

From the first of these one sees that

$$X_1(z) = e^{cz}P(z),$$

where $P(z)$ is periodic. If one lets

$$X_2(z) = e^{cz}Q(z),$$

one finds from the second equation in (13) that

$$Q(z + T) - Q(z) = \frac{1}{\mu} P(z).$$

Clearly, a solution of the above equation is

$$Q(z) = \frac{z}{\mu T} P(z).$$

Similarly, it follows from the remaining equations that

$$X_j(z) = \frac{z(z - T)(z - 2T) \cdots (z - (j - 2)T)}{(j - 1)!\,(\mu T)^{j-1}} e^{cz}P(z). \tag{14}$$

From this discussion one can derive the result that the general solution of (5) must have the form

$$X = \sum_{i=1}^{n} a_i e^{c_i z} p_i(z) P_i(z), \tag{15}$$

where the a_i are arbitrary scalars, the c_i are suitable exponents, the $p_i(z)$ are polynomials, and the $P_i(z)$ are periodic vectors. These results are essentially due to Floquet, and the general approach is often referred to as Floquet theory.

5.2 APPLICATIONS TO SECOND-ORDER EQUATIONS

We now return to the equation

$$y'' + [\lambda - q(z)]y = 0, \tag{1}$$

which was treated in the preceding chapter, but we make the additional stipulation that $q(z)$ have period π; that is,

$$q(z + \pi) = q(z).$$

For the moment λ will represent a constant parameter.

To apply the methods of the preceding section, we let

$$y = x_1,$$
$$y' = x_2,$$

so that (1) can be rewritten as the system

$$X' = \begin{pmatrix} 0 & 1 \\ q(z) - \lambda & 0 \end{pmatrix} X. \tag{2}$$

If $\Phi(z)$ is a fundamental matrix solution of (2) such that

$$\Phi(0) = I,$$

then as before we have

$$\Phi(z + \pi) = \Phi(z)\Phi(\pi).$$

We now introduce two solutions of (1) with the initial values

$$\begin{aligned} y_1(0) &= 1, \qquad y_1'(0) = 0, \\ y_2(0) &= 0, \qquad y_2'(0) = 1, \end{aligned} \tag{3}$$

so that

$$\Phi(z) = \begin{pmatrix} y_1 & y_2 \\ y_1' & y_2' \end{pmatrix}.$$

Next we seek the eigenvalues of $\Phi(\pi)$:

$$\begin{aligned} | \Phi(\pi) - \mu I | &= \begin{vmatrix} y_1(\pi) - \mu & y_2(\pi) \\ y_1'(\pi) & y_2'(\pi) - \mu \end{vmatrix} \\ &= \mu^2 - (y_1(\pi) + y_2'(\pi))\,\mu + (y_1(\pi)\,y_2'(\pi) - y_2(\pi)\,y_1'(\pi)) \\ &= \mu^2 - (y_1(\pi) + y_2'(\pi))\,\mu + 1. \end{aligned} \tag{4}$$

The latter follows from the fact that

$$y_1(\pi)\,y_2'(\pi) - y_2(\pi)\,y_1'(\pi) = | \Phi(\pi) | = | \Phi(0) | = 1.$$

We conclude therefore that it must be possible to find solutions of (2) such that

$$\begin{aligned} X_1(z+\pi) &= \mu_1 X_1(z), \\ X_2(z+\pi) &= \mu_2 X_2(z), \end{aligned} \tag{5}$$

where μ_1 and μ_2 represent the zeros of (4), provided they are distinct. From these we can extract solutions of (1) with similar properties. In general these solutions will not be periodic. But we can deduce even more information from (4); namely, if

$$| y_1(\pi) + y_2'(\pi) | < 2,$$

μ_1 and μ_2 will be complex conjugate zeros of (4), and since

$$\mu_1 \mu_2 = 1,$$

it follows that

$$\mu_1 = e^{i\theta}, \qquad \mu_2 = e^{-i\theta}$$

for a suitable angle θ. In this case all solutions of (1) must be bounded, since if w_i is a solution of (1) such that

$$w_i(z+\pi) = \mu_i w_i(z), \qquad i = 1, 2,$$

then

$$| w_i(z+\pi) | = | w_i(z) |,$$

which shows that $w_i(z)$ is bounded. Therefore any linear combination of $w_1(z)$ and $w_2(z)$ is also bounded.

When, however,

$$| y_1(\pi) + y_2'(\pi) | > 2,$$

we find that the zeros of (4) are real, and since

$$\mu_1 \mu_2 = 1,$$

then one of these (say, μ_1) must be such that

$$| \mu_1 | > 1,$$

and

$$| \mu_2 | = \left| \frac{1}{\mu_1} \right| < 1.$$

Then from

$$w_1(z+\pi) = \mu_1 w_1(z),$$

we see that

$$|w_1(z+\pi)| = |\mu_1|\,|w_1(z)|,$$

and more generally,

$$|w_1(z+k\pi)| = |\mu_1|^k\,|w_1(z)|.$$

From this we see that as k increases beyond bound, so does $|w_1(z+k\pi)|$. We call the former case the stable case and the latter the unstable one.

The situation where

$$|y_1(\pi) + y_2'(\pi)| = 2$$

still has to be discussed. When

$$y_1(\pi) + y_2'(\pi) = 2,$$

we see that (4) has a double zero at $\mu = 1$. In this case we can certainly find a solution of period π, since

$$w_1(z+\pi) = \mu_1 w_1(z) = w_1(z).$$

As shown in the preceding section, the second solution $w_2(z)$ either has also period π or is of the form $zw_1(z)$.

In the remaining case where

$$y_1(\pi) + y_2'(\pi) = -2$$

a similar situation arises. The equation (4) has a double zero at $\mu = -1$, and we find at least one solution such that

$$w_1(z+\pi) = -w_1(z),$$

and therefore

$$w_1(z+2\pi) = -w_1(z+\pi) = w_1(z),$$

so that $w_1(z)$ has period 2π. The second solution again has period 2π or is of the form $zw_1(z)$.

Until now λ has been held fixed. In this case the solutions of (1) will depend on λ. We let

$$\Delta(\lambda) = y_1(\pi) + y_2'(\pi)$$

and call $\Delta(\lambda)$ the discriminant of (1). To find values of λ for which (1) has solutions of period π, we must solve the equation

$$\Delta(\lambda) = 2. \tag{6}$$

To find the values of λ for which (1) has solutions of period 2π, we must solve

$$\Delta(\lambda) = -2; \tag{7}$$

and to obtain both sets of λ's, we solve

$$\Delta^2(\lambda) = 4. \tag{8}$$

EXAMPLE: Consider the equation

$$y'' + \lambda y = 0.$$

The solutions that satisfy the initial conditions (3) are given by

$$y_1 = \cos \sqrt{\lambda}\, z,$$

$$y_2 = \frac{\sin \sqrt{\lambda}\, z}{\sqrt{\lambda}},$$

so that

$$\Delta(\lambda) = y_1(\pi) + y_2'(\pi) = 2 \cos \sqrt{\lambda}\, \pi.$$

The solutions of (6) are given by

$$\sqrt{\lambda} = 2n, \qquad n = 0, 1, 2, \cdots$$

and the solutions of (7) by

$$\sqrt{\lambda} = 2n + 1, \qquad n = 0, 1, 2, \cdots.$$

The solutions of both sets, that is, the solutions of (8), are given by

$$\sqrt{\lambda} = n, \qquad n = 0, 1, 2, \cdots.$$

When $\sqrt{\lambda} = 0$, we have only one periodic solution, namely,

$$y_1 = 1.$$

However, for all the other λ's we obtain two periodic solutions. Thus

$$y'' + 4n^2 y = 0, \qquad n = 1, 2, \cdots$$

has two solutions of period π, namely,

$$y_1 = \cos 2nz,$$

$$y_2 = \frac{\sin 2nz}{2n},$$

and for $\sqrt{\lambda} = 2n + 1$, we find

$$y_1 = \cos(2n+1)z,$$

$$y_2 = \frac{\sin(2n+1)z}{2n+1},$$

which have period 2π.

The solutions of (6) could be characterized as yielding the eigenvalues of (1) subject to the boundary conditions

$$\begin{aligned} y(0) &= y(\pi), \\ y'(0) &= y'(\pi). \end{aligned} \tag{9}$$

To see that these boundary conditions yield periodic solutions, and conversely, that the roots of (6) yield solutions satisfying (9), one returns to the system concept. Then the solution of (2) can be written in the form

$$X(z) = \Phi(z)X(0), \tag{10}$$

where $X(0)$ is the initial vector. Evidently

$$X(\pi) = \Phi(\pi)X(0).$$

If $\Delta(\lambda) = 2$, $\Phi(\pi)$ must have an eigenvalue $\mu = 1$. In this case the equation

$$\Phi(\pi)X(0) = X(0) \tag{11}$$

has a nonzero solution, and it follows that

$$X(\pi) = X(0).$$

Conversely, if the latter holds from (9), then (11) has nonzero solutions and $\Phi(\pi)$ must have $\mu = 1$ as an eigenvalue.

Similarly one can show that the solutions of (7) can be characterized as yielding the eigenvalues of (1) subject to the boundary conditions

$$\begin{aligned} y(0) &= -y(\pi), \\ y'(0) &= -y'(\pi). \end{aligned} \tag{12}$$

If the eigenvalues corresponding to the boundary conditions (9) are denoted by $\lambda_0, \lambda_1, \lambda_2, \cdots$, and those corresponding to (12) by $\lambda_1', \lambda_2', \lambda_3', \cdots$, it can be shown that

$$\lambda_0 < \lambda_1' \leqslant \lambda_2' < \lambda_1 \leq \lambda_2 < \lambda_3' \leq \lambda_4' < \lambda_3 \leq \lambda_4 < \cdots. \tag{13}$$

The proof of this statement is very complicated and will be omitted. For the special example discussed in this section, (13) can be checked explicitly. One can then show that if the discriminant $\Delta(\lambda)$ is plotted as a function of λ, one obtains the graph given in Fig. 5.1.

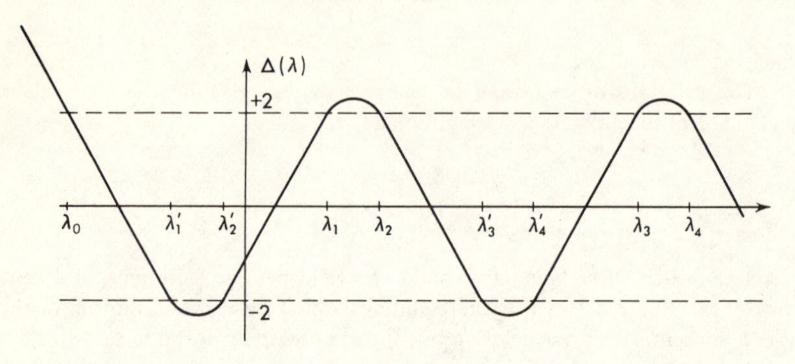

Fig. 5.1

From Fig. 5.1 it is evident that for values of λ in the intervals $(-\infty, \lambda_0)$, (λ_1', λ_2'), (λ_1, λ_2), (λ_3', λ_4'), $(\lambda_3, \lambda_4) \cdots$, $|\Delta(\lambda)| > 2$, and one solution of (1) is unbounded. These intervals are known as the instability intervals. For values of λ in (λ_0, λ_1'), (λ_2', λ_1), (λ_2, λ_3'), (λ_4', λ_3), $\cdots$, $|\Delta(\lambda)| < 2$ and all solutions of (1) are bounded. These intervals are known as the stability intervals.

Problems

1. If $\Phi(z)$ is an arbitrary fundamental matrix solution of

$$\Phi' = A(z)\Phi,$$

where A has period T, show that the roots of

$$|\Phi(T) - \mu\Phi(0)| = 0$$

are independent of the fundamental matrix selected.

Hint: Show that all fundamental matrix solutions can be expressed in terms of one satisfying (12).

2. The following is a sketch of a proof that the periodic system

$$X' = A(z)X$$

can, by a substitution of variables, be reduced to one with constant coefficients.

Let $\Phi(z)$ be a fundamental matrix solution. Then there must exist a constant matrix C such that

$$\Phi(z + T) = \Phi(z)C.$$

Furthermore there must exist, but not in a unique way, a matrix B such that

$$e^{TB} = C.$$

Let

$$X = \Phi e^{-zB} Y,$$

so that

$$\begin{aligned} X' &= \Phi' e^{-zB} Y - \Phi B e^{-zB} Y + \Phi e^{-zB} Y' = A(z)X \\ &= A\Phi e^{-zB} Y. \end{aligned}$$

It follows that

$$Y' = BY.$$

Hint: Carry out the details of the proof. Use (12) and the fact that no diagonal element in J can vanish.

3. Consider the periodic system

$$X' = [A + B(z)]X,$$

where A is a constant matrix with eigenvalues whose real parts are negative, and $B(z)$ has period T. Show that if $||B(z)||$ is sufficiently small, for all solutions

$$\lim_{z \to \infty} X = 0.$$

Hint: Compare with the stability theorems of Chapter 6.

4. In the equation

$$y'' + [\lambda - q(z)]y = 0,$$

one can assume without loss of generality that

$$\int_0^{\pi} q(z)\, dz = 0;$$

otherwise, by a slight redefinition of λ and q, this can be achieved. Show that λ_0, the first eigenvalue corresponding to a solution of period π must be nonpositive if the mean value of $q(z)$ vanishes.

To prove the result, you must first prove that the eigenfunction y_0 has no zeros.

5. In many physical problems $q(z)$ is an even function of z; that is, $q(z) = q(-z)$. Show that if that is the case, then if periodic solutions exist, it must be possible to find either even or odd solutions. The only case in which one could find periodic solutions that are not even or odd would arise in the case of a double eigenvalue in which case all solutions would be periodic.

6. When $q(z)$ is even show that the eigenfunctions and eigenvalues could be characterized in terms of boundary value problems with separated end point conditions, as follows:

(a) $y'(0) = y'\left(\frac{\pi}{2}\right) = 0$ (even solutions of period π).

(b) $y(0) = y\left(\frac{\pi}{2}\right) = 0$ (odd solutions of period π).

(c) $y'(0) = y\left(\frac{\pi}{2}\right) = 0$ (even solutions of period 2π).

(d) $y(0) = y'\left(\frac{\pi}{2}\right) = 0$ (odd solutions of period 2π).

7. Consider the following differential equation:

$$y'' + \lambda^2 q(z) y = 0$$

where

$$\begin{aligned} q(z) &= 1, \qquad 0 \leqslant z < 1, \\ &= a^2, \qquad 1 \leqslant z \leqslant L, \end{aligned}$$

and $q(z)$ is to be continued as an even periodic function outside the interval $(0, L)$.

(a) Use the results of the preceding problem to reduce the problem to a study of four boundary value problems, with separated boundary conditions.

(b) Study the solutions of the four transcendental equations obtained in (a) and show that the eigenvalues satisfy (13) of Section 5.2.

(c) Show that equalities in (13) can arise if and only if $a(L - 1)$ is a rational number.

8. As in problem 14 of Chapter 4, show that the smallest eigenvalue of

$$y'' + \lambda q(z) y = 0,$$

$$y(0) = -y(\pi), \qquad y'(0) = -y'(\pi),$$

where $q(z) \geqslant 0$ and integrable, is greater than unity provided

$$\pi \int_0^\pi q(z)\, dz \leq 4.$$

9. Use the results of the preceding exercise to show that all solutions of

$$y'' + q(z)y = 0,$$

where $q(z + \pi) = q(z)$, $q(z) \geqslant 0$ and $\pi \int_0^\pi q dz \leqslant 4$.

10. Consider

$$y'' + p(z)y' + a^2 y = 0,$$

where $p(z + \pi) = p(z) > 0$, and a is constant. Show that for all solutions

$$\lim_{z \to \infty} y = 0.$$

6 Nonlinear Differential Equations

6.1 FUNDAMENTAL EXISTENCE THEOREM

In this chapter we turn our attention to nonlinear differential equations. These can in general be written in the form of systems, as in the linear case, as follows:

$$x_i' = f_i(x_1, x_2, \cdots, x_n, z), \qquad i = 1, 2, \cdots, n. \tag{1}$$

It will be convenient to introduce a vector formalism by defining

$$X = \begin{pmatrix} x_1 \\ x_2 \\ \vdots \\ x_n \end{pmatrix}, \qquad F(X, z) = \begin{pmatrix} f_1(x_1, \cdots, x_n, z) \\ f_2(x_1, \cdots, x_n, z) \\ \vdots \\ f_n(x_1, \cdots, x_n, z) \end{pmatrix}$$

and rewriting (1) in the form

$$X' = F(X, z). \tag{2}$$

We shall suppose that a set of initial conditions is prescribed, say,

$$X(0) = X_0,$$

and we shall then show that under suitable restrictions on $F(X, z)$, (2) has a unique solution. The method of proof is completely analogous to the corresponding proof for linear systems given in Section 2.8; that is, we shall replace (2) by the integral equation

$$X = X_0 + \int_0^z F(X, t)\, dt$$

and then show that the successive approximations obtained from

$$X_{n+1} = X_0 + \int_0^z F(X_n, t)\, dt, \qquad n = 0, 1, 2, \cdots \tag{3}$$

converge to the solution of (2) as n approaches infinity.

The function $F(X, z)$ will satisfy a so-called Lipschitz condition, provided a constant k can be found such that

$$\| F(X_1, z) - F(X_2, z) \| \leqslant k \| X_1 - X_2 \|$$

for all X_1 and X_2 in the domain

$$\| X - X_0 \| \leqslant a$$

and all z in the interval

$$| z | \leqslant b.$$

THEOREM: The differential equation

$$X' = F(X, z)$$

with initial value

$$X(0) = X_0$$

has a unique solution for

$$\| X - X_0 \| \leqslant \min (a, b/M)$$
$$| z | \leqslant b,$$

provided $F(X, z)$ satisfies a Lipschitz condition in that domain and is also bounded so that

$$\| F(X, z) \| \leqslant M.$$

To prove the theorem, we return to (3) and see that

$$X_{n+1} - X_n = \int_0^z [F(X_n, t) - F(X_{n-1}, t)]\, dt.$$

From the above and use of the Lipschitz condition we find that

$$\| X_{n+1} - X_n \| \leqslant \int_0^z \| F(X_n, t) - F(X_{n-1}, t) \|\, dt$$
$$\leqslant k \int_0^z \| X_n - X_{n-1} \|\, dt.$$

From this point on, the proof proceeds as in Section 2.8. One can now show by mathematical induction that

$$\| X_{n+1} - X_n \| \leqslant \frac{Mk^n z^{n+1}}{(n+1)!}.$$

From the identity

$$X_n = X_0 + (X_1 - X_2) + \cdots + (X_n - X_{n-1}),$$

it follows that

$$\| X_n \| \leqslant \| X_0 \| + \sum_0^\infty \frac{Mk^n z^{n+1}}{(n+1)!} = \| X_0 \| + \frac{M}{k} (e^{kz} - 1).$$

It is evident that

$$X = X_0 + \lim_{n\to\infty} \sum_{i=1}^{n} (X_i - X_{i-1}) = X_0 + \sum_{i=1}^{\infty} (X_i - X_{i-1})$$

exists, since the series on the right converges. It follows that the solutions of (3) converge to a solution of (2).

To complete the uniqueness proof, we now assume that two solutions exist, say, X and Y. Then

$$X = X_0 + \int_0^z F(X, t)\, dt,$$

$$Y = X_0 + \int_0^z F(Y, t)\, dt,$$

so that

$$\| X - Y \| \leqslant \int_0^z \| F(X, t) - F(Y, t) \| \, dt$$

$$\leqslant k \int_0^z \| X - Y \| \, dt.$$

By successive integrations of the last inequality, it follows that

$$\| X - Y \| \leqslant \frac{k^{n+1}}{n!} \int_0^z (z - t)^n \| X - Y \| \, dt,$$

and as n tends to infinity, we see that

$$\| X - Y \| = 0,$$

so that

$$X \equiv Y.$$

This completes the uniqueness proof.

By a similar argument one can show that the solution depends continuously on the initial data. Suppose X and Y are solutions of (2) corresponding to the initial conditions X_0 and Y_0 where

$$\| X_0 - Y_0 \| \leqslant \delta.$$

Then, as above,

$$\| X - Y \| \leqslant \delta + k \int_0^z \| X - Y \| \, dt,$$

and one can show that

$$\| X - Y \| \leqslant \delta e^{kz}.$$

The importance of the conditions of the theorem is seen by examining the equation

$$y' = \frac{2y}{z}$$

which was discussed in Section 2.2. There it is shown that the equation has nonunique solutions in the neighborhood of the point (0, 0), where the right side is unbounded.

6.2 SERIES SOLUTIONS

The existence and uniqueness theorem of the preceding section is of great theoretical importance, but is hardly of much use in obtaining some information about the solution of the differential equation. The Lipschitz condition,

$$\| F(X_1, z) - F(X_2, z) \| \leqslant k \| X_1 - X_2 \|,$$

is a very general condition. One can easily show that if the components of $F(X, z)$ are bounded and possess bounded first partial derivatives with respect to the dependent variables x_i, then the Lipschitz condition has been satisfied. But in many practical cases the components of $F(X, z)$ are analytic; that is, they can be expanded into power series in the independent variable and the dependent variables.

In these cases it is possible to show that the solution of the differential equation also can be represented by a power series, which converges in the neighborhood of the initial condition. To prove this result, we first require a preliminary lemma.

LEMMA. If the series

$$f(z) = \sum_0^\infty c_n z^n$$

converges absolutely for all $|z| \leqslant r$, then there exists another series

$$g(z) = \sum_0^\infty d_n z^n,$$

which also converges absolutely for $|z| \leqslant r' < r$ and

$$d_n \geqslant |c_n|, \qquad n = 0, 1, 2, 3, \cdots.$$

Since $\sum_0^\infty |c_n| r^n$ converges by hypotheses, we see that

$$\lim_{n\to\infty} |c_n| r^n = 0.$$

Therefore there must exist a positive constant M such that

$$| c_n | r^n \leqslant M.$$

If we now let

$$d_n = \frac{M}{r^n},$$

we find that

$$g(z) = \sum_0^\infty \frac{M}{r^n} z^n = \frac{M}{1 - \frac{z}{r}},$$

which converges for all $| z | \leqslant r' < r$, as required by the lemma.

The series $g(z)$ will be said to dominate $f(z)$. It should be observed that the r used may not be the radius of convergence of $f(z)$, but any smaller quantity will do. The above lemma can easily be extended to a power series in n variables. Then, if

$$f(z) = \sum_{k_i \geqslant 0} c_{k_1, k_2, \cdots k_n} z_1^{k_1} z_2^{k_2} \cdots z_n^{k_n}$$

and if this series converges absolutely for

$$| z_i | \leqslant r_i,$$

it must be possible to find positive constants M and r such that

$$r < r_i, \qquad i = 1, 2, \cdots n,$$

and

$$g(z) = \sum_{k_i \geqslant 0} M \left(\frac{z_1}{r}\right)^{k_1} \left(\frac{z_2}{r}\right)^{k_2} \cdots \left(\frac{z_n}{r}\right)^{k_n}$$

$$= \frac{M}{\left(1 - \frac{z_1}{r}\right)\left(1 - \frac{z_2}{r}\right) \cdots \left(1 - \frac{z_n}{r}\right)},$$

where

$$\frac{M}{r^n} = \frac{M}{r^{k_1 + k_2 + \ldots + k_n}} \geqslant | c_{k_1, k_2, \ldots, k_n} |.$$

We are now in a position to state and prove the following theorem.

THEOREM: The differential equation

$$x' = f(x, z), \tag{1}$$

where $f(x, z)$ is analytic in the neighborhood of the initial point

$$x(z_0) = x_0,$$

has an analytic solution

$$x = x_0 + \sum_{1}^{\infty} a_n(z - z_0)^n. \tag{2}$$

In the first place we can assume without loss of generality that $x_0 = z_0 = 0$. This can always be accomplished by a translation of coordinates. Next we rewrite (1) in the form

$$x' = \sum_{k_i \geq 0} c_{k_1,k_2} x^{k_1} z^{k_2}$$

and then replace x by the series (2). The unknown coefficients a_n will then be determined by comparing powers of z.

$$\sum_{1}^{\infty} n a_n z^{n-1} = \sum_{k_i \geq 0} c_{k_1,k_2} \left(\sum_{1}^{\infty} a_n z^n\right)^{k_1} z^{k_2}$$

$$= c_{0,0} + (c_{1,0}a_1 + c_{0,1})\, z + (c_{2,0}a_1^2 + c_{1,1}a_1 + c_{0,2} + c_{1,0}a_2)\, z^2 + \cdots. \tag{3}$$

Then

$$\begin{aligned} a_1 &= c_{0,0}, \\ 2a_2 &= c_{1,0}a_1 + c_{0,1}, \\ 3a_3 &= c_{2,0}a_1^2 + c_{1,1}a_1 + c_{0,2} + c_{1,0}a_2 \\ &\vdots \\ na_n &= p_n(c_{k_1,k_2}, a_1, a_2, \cdots, a_{n-1}) \\ &\vdots \end{aligned} \tag{4}$$

where the p_n are polynomials in the c_{k_1,k_2} and $a_1, a_2, \cdots, a_{n-1}$. From these recursion formulas the a_n can be determined, and it is clear that each of the a_n is expressed in terms of a polynomial involving the coefficients c_{k_1,k_2} and the preceding a_k. The next task is to show that the resultant series (2) does indeed converge. To do so, we first examine the equation

$$x' = g(x, z) = \frac{M}{\left(1 - \frac{x}{r}\right)\left(1 - \frac{z}{r}\right)}, \tag{5}$$

$$x(0) = 0,$$

where $g(x, z)$ dominates $f(x, z)$. This equation can be solved very easily. An integration leads to

$$-\frac{r}{2}\left(1-\frac{x}{r}\right)^2+\frac{r}{2}=-rMln\left(1-\frac{z}{r}\right),$$

and solving for x, we find

$$x=r\left[1-\left(1+2Mln\left(1-\frac{z}{r}\right)\right)^{1/2}\right]. \tag{6}$$

The negative square root is selected here to ensure that

$$x(0)=0.$$

(6) is an analytic function and therefore has a series expansion:

$$x=\sum_1^\infty d_n z^n. \tag{7}$$

The convergence of this series depends on the singularities of the solution (6). The nearest singularity to the origin will lie at the point where the radical vanishes; that is,

$$1+2Mln\left(1-\frac{z}{r}\right)=0,$$

so that for convergence, we require that

$$|z|<r\left(1-\exp\frac{-1}{2M}\right). \tag{8}$$

Had we attempted to determine the d_n in (7) directly from (5), we should have been led to a set of recursion formulas:

$$nd_n=p_n\left(\frac{M}{r^{k_1+k_2}},d_1,d_2,\cdots,d_{n-1}\right). \tag{9}$$

The polynomials in (9) are identical to those in (4), but the arguments are different. Clearly,

$$|a_1|=|c_{0,0}|\leqslant M=d_1$$

$$2|a_2|\leqslant|c_{1,0}||a_1|+|c_{0,1}|\leqslant\frac{M}{r}d_1+\frac{M}{r}=2d_2,$$

and generally

$$n|a_n|\leqslant|p_n(c_{k_1,k_2},a_1,a_2,\cdots,a_{n-1})|$$

$$\leqslant p_n\left(\frac{M}{r^{k_1+k_2}},d_1,d_2,\cdots,d_{n-1}\right)=nd_n$$

Therefore (7) dominates (2) and (2) also converges in the domain (8).

In the case of linear differential equations with analytic coefficients it was shown in Section 3.1 that the series solutions converged in every domain in which the coefficients were analytic. That this need no longer be true in the case of nonlinear equations can easily be demonstrated with the following example:

$$x' = 1 + x^2, \qquad x(0) = 0. \tag{10}$$

The theorem clearly applies to this case, and a series solution can be found. An explicit solution is given by

$$x = \tan z,$$

which has a singularity at $z = \pi/2$. But the right side of (10) is analytic for all finite x and z.

The preceding theorem can be easily extended to first-order systems of equation.

THEOREM: The differential equation

$$X' = F(X, z), \qquad X(0) = 0, \tag{11}$$

where the components of $F(X, z)$ are analytic, has an analytic solution

$$X = \sum_{1}^{\infty} A_n z^n. \tag{12}$$

The proof is completely analogous to that of the preceding case. First, all components of $F(X, z)$ are expanded in series:

$$f_i(x_1, x_2, \cdots, x_n, z) = \sum_{k_j \geqslant 0} c^{(i)}_{k_1 k_2, \cdots k_{n+1}} x_1^{k_1} x_2^{k_2} \cdots x_n^{k_n} z^{k_{n+1}}$$

$$i = 1, 2, \cdots n.$$

Next a solution of the form

$$x_i = \sum_{1}^{\infty} a_k^{(i)} z^k$$

is assumed and inserted in (11). This leads to recursion formulas of the form

$$k a_k^{(i)} = p_k(c^{(j)}_{k_1, k_2, \cdots k_{n+1}}, a_1^{(j)}, a_2^{(j)}, \cdots, a_{k-1}^{(j)}), \tag{13}$$

where the p_k are polynomials in the coefficients of all f_i and all $a_l^{(j)}$ with $l < k$, $1 \leqslant j \leqslant n$.

As before, all f_i are dominated by a single function:

$$g(x_1, x_2, \cdots, x_n, z) = \frac{M}{\left(1 - \frac{x_1}{r}\right)\left(1 - \frac{x_2}{r}\right)\cdots\left(1 - \frac{x_n}{r}\right)\left(1 - \frac{z}{r}\right)}.$$

Instead of (11) we consider the system

$$\begin{aligned} x_i' &= g(x_1, x_2, \cdots, x_n, z), \qquad i = 1, 2, \cdots, n, \\ x_i(0) &= 0. \end{aligned} \tag{14}$$

By symmetry it is clear that for (14) we must have

$$x_1 = x_2 = \cdots = x_n.$$

Then (14) reduces to a single equation

$$x' = \frac{M}{\left(1 - \frac{x}{r}\right)^n\left(1 - \frac{z}{r}\right)},$$

$$x(0) = 0,$$

whose solution is given by

$$x = r\left[1 - \left(1 + M(n+1)\, ln\left(1 - \frac{z}{r}\right)\right)^{1/(n+1)}\right].$$

The preceding series will converge for all z for which the expression in the radical is positive; that is,

$$|z| < r\left(1 - \exp\left(\frac{-1}{M(n+1)}\right)\right).$$

If we let

$$x = \sum_1^\infty d_n z^n,$$

we find as before that

$$k\,|a_k^{(i)}| \leqslant kd_k,$$

so that (12) converges.

Another important type of series solution arises in cases where the differential equation involves a parameter. For example, we have the following theorem.

THEOREM: If $F(X, z, \mu)$ is analytic in $x_1, x_2, \cdots, x_n$ and μ, and continuous in z, the system

$$X' = F(X, z, \mu), \quad X(0) = 0, \tag{15}$$

will, have a solution of the form

$$X = \sum_0^\infty X_k(z)\, \mu^k,$$

where the $X_k(z)$ are suitable vector functions of z.

To prove the theorem, we shall first consider a simple case in which we have only one dependent variable:

$$x' = f(x, z, \mu), \quad x(0) = 0. \tag{16}$$

For $\mu = 0$, we have

$$x' = f(x, z, o),$$
$$x(o) = 0,$$

and we shall call the solution of this equation $x_0(z)$. We now return to (16) and let

$$x = x_0(z) + y(z),$$

so that we obtain

$$y'(z) = f(x_0(z) + y(z), z, \mu) - x_0'(z) \equiv g(y, z, \mu) \tag{17}$$

Clearly this equation has, for $\mu = 0$, the solution $y = 0$. We can now expand $g(y, z, \mu)$ in a power series in y and μ, where the coefficients are continuous functions of z.

$$\begin{aligned} y' &= \sum_{k_i \geqslant 0} c_{k_1, k_2}(z)\, y^{k_1} \mu^{k_2} \\ &= c_{1,0}(z)\, y + c_{0,1}(z)\, \mu \\ &\qquad + c_{2,0}(z)\, y^2 + c_{1,1}(z)\, y\mu + c_{0,2}^{(z)}\, \mu^2 + \cdots, \\ y(0) &= 0, \end{aligned} \tag{18}$$

where $c_{0,0}(z)$ has been set equal to zero, since for $\mu = 0$, the solution is $y = 0$. We now seek a solution in the form

$$y = \sum_1^\infty y_n(z)\, \mu^n \tag{19}$$

and find that

$$
\begin{aligned}
y_1' &= c_{1,0}(z)\, y_1 + c_{0,1}(z), \\
y_2' &= c_{1,0}(z)\, y_2 + c_{2,0}(z)\, y_1^2 + c_{1,1}(z)\, y_1 + c_{0,2}(z), \\
&\vdots \\
y_n' &= c_{1,0}(z)\, y_n + p_n(c_{k_1,k_2}(z), y_1, y_2, \cdots, y_{n-1}), \\
y_n(0) &= 0, \qquad n = 1, 2, 3, \cdots.
\end{aligned}
\tag{20}
$$

The infinite system (20) is a set of linear differential equations that can be solved recursively for the y_n. We must now show that the series (19) does converge. To do so, we seek a function of y and μ whose series expansion dominates the series in (18). If the series in (18) converges for all $|y| \leqslant r$ and $|\mu| \leqslant r$ and has a maximum of M for those values of y and μ, then a suitable dominating function is

$$
\frac{M\left(\dfrac{y+\mu}{r}\right)\left(1 + \dfrac{y+\mu}{r}\right)}{1 - \dfrac{y+\mu}{r}}.
$$

Such dominating functions are not unique, but can be selected in numerous ways. The ultimate choice is dictated by convenience. The determining factors are whether or not they lead to readily integrable differential equations and whether the resultant convergence criteria are adequate.

We now consider the differential equation

$$
\begin{aligned}
y' &= \frac{M\left(\dfrac{y+\mu}{r}\right)\left(1 + \dfrac{y+\mu}{r}\right)}{1 - \dfrac{y+\mu}{r}}, \\
y(0) &= 0,
\end{aligned}
\tag{21}
$$

and find that its solution is given by

$$
\frac{\dfrac{y+\mu}{r}}{\left(1 + \dfrac{y+\mu}{r}\right)^2} = \frac{\dfrac{\mu}{r}}{\left(1 + \dfrac{\mu}{r}\right)^2}\, e^{(M/r)z},
\tag{22}
$$

from which we obtain

$$
\frac{y+\mu}{r} = Q - 1 - \sqrt{Q^2 - 2Q},
\tag{23}
$$

$$
Q = \frac{\left(1 + \dfrac{\mu}{r}\right)^2}{\dfrac{2\mu}{r}}\, e^{-(M/r)z}.
$$

Now (23) can be expanded in a power series in the variable μ, and this series will converge for all μ for which the radical in (23) is positive. Therefore we require that

$$|Q| > 2,$$

which in turn leads to

$$|\mu| < r\frac{1-\sqrt{1-e^{-Mz/r}}}{1+\sqrt{1-e^{-Mz/r}}}. \tag{24}$$

If we seek a solution for (21) in the form

$$y = \sum_{1}^{\infty} V_n(z)\,\mu^n,$$

we obtain, as in (20),

$$\begin{aligned} V_n' &= d_{1,0}V_n + p_n(d_{k_1,k_2}, V_1, V_2, \cdots, V_{n-1}), \\ V_n(0) &= 0, \qquad n = 0, 1, 2, \cdots. \end{aligned} \tag{25}$$

From (20) and (25) we find that

$$y_n = \int_0^z p_n(c_{k_1,k_2}(z), y_1, y_2, \cdots, y_{n-1}) \exp\left(\int_t^z c_{1,0}(t)dt\right) dt$$

$$V_n = \int_0^z p_n(d_{k_1,k_2}, V_1, V_2, \cdots, V_{n-1}) \exp\,(d_{1,0}(z-t)\,dt),$$

and clearly the latter of these dominates the former and it follows that

$$|y_n| \leqslant V_n.$$

Therefore (19) converges, at least for those values of μ for which (24) holds. It may even converge for larger μ domains, but a different converge proof would have to be used.

The proof for the case of nth order systems does not differ in any significant way from the preceding proof. We return to (15) and first solve the system for $\mu = 0$:

$$\begin{aligned} X' &= F(X, z, 0), \\ X(0) &= 0. \end{aligned}$$

The solution of the above will be called $X_0(z)$, and we now let

$$X = X_0(z) + Y(z)$$

so that (15) is replaced by

$$\begin{aligned} Y' &= F(X_0(z) + Y(z), z, \mu) - X_0'(z) \equiv G(Y, z, \mu), \\ Y(0) &= 0. \end{aligned} \tag{26}$$

Analogous to (17), (26) has, for $\mu = 0$, the solution $Y = 0$. As in (18) the system (26) can be expanded in power series so that the equation for the l'th component of Y becomes, as in (18),

$$\begin{aligned} y^{(l)\prime} &= \sum_{k_i \geqslant 0} c_{k_1, k_2, \cdots, (z)_{k_{n+1}}} (y^{(1)})^{k_1} \cdots (y^{(n)})^{k_n} \mu^{k_{n+1}} \\ &= c_{1,0,0\cdots 0}(z)\, y^{(1)} + c_{0,2,0\cdots 0}(z)\, y^{(2)} + \cdots + c_{0,0,\cdots ,0,1,0}(z)\, y^{(n)} \\ &\quad + c_{0,0,\cdots ,0,1}(z)\, \mu + \cdots \text{(higher order terms)}, \end{aligned} \tag{27}$$

$$y^{(l)}(0) = 0,$$
$$l = 1, 2, \cdots, n.$$

In view of the fact that $y^{(l)} = 0$ is the solution for $\mu = 0$, we require that

$$c_{0,0,0,\cdots 0}(z) = 0.$$

Next we seek a solution for (27) in the form

$$y^{(l)} = \sum_{m=1}^{\infty} y_m^{(l)}(z)\, \mu^m, \qquad l = 1, 2, \cdots, n, \tag{28}$$

and find, as in (20),

$$\begin{aligned} y_m^{(l)\prime} &= c_{1,0,\cdots ,0}(z)\, y_m^{(1)} + c_{0,1,0,\cdots 0}(z)\, y_m^{(2)} + \cdots + c_{0,0,\cdots ,0,\cdots ,0,1,0}(z)\, y_m^{(n)} \\ &\quad + p_m^{(l)}(c(z), y_1^{(k)}, y_2^{(k)}, \cdots, y_{m-1}^{(k)}), \end{aligned} \tag{29}$$

$$y_m^{(l)}(0) = 0, \qquad l = 1, 2, \cdots, n, \quad m = 1, 2, 3, \cdots.$$

The term $p_m^{(l)}$ is a polynomial in the coefficients and terms like $y_j^{(k)}$, where the lower index $j < m$; (29) represents linear systems of equations from which the $y_m^{(l)}$ can be determined recursively. If the series in (28) can be shown to converge, then a solution for (26) will have been found. To do so, we first require a dominating function for the right side of (27). The resulting system of differential equations, analogous to (21), is

$$y^{(l)\prime} = \frac{M \left(\dfrac{y^{(1)} + y^{(2)} + \cdots + y^{(n)} + \mu}{r} \right) \left(1 + \dfrac{y^{(1)} + y^{(2)} + \cdots + y^{(n)} + \mu}{r} \right)}{1 - \dfrac{y^{(1)} + y^{(2)} + \cdots + y^{(n)} + \mu}{r}} \tag{30}$$

$$y^{(l)}(0) = 0, \qquad l = 1, 2, \cdots, n.$$

One can easily show that the right side of (30) dominates the right side of (27). The solutions of (30), by the symmetry of the system, can easily be shown to satisfy

$$y^{(1)} = y^{(2)} = \cdots y^{(n)}.$$

We call all of these y, replacing (30) by

$$y' = \frac{M\left(\frac{ny+\mu}{r}\right)\left(1+\frac{ny+\mu}{r}\right)}{1-\frac{ny+\mu}{r}},$$

$$y(0) = 0.$$

Analogous to (23) we find

$$\frac{ny+\mu}{r} = Q - 1 - \sqrt{Q^2 - 2Q}$$

$$Q = \frac{\left(1+\frac{\mu}{r}\right)^2}{\frac{2\mu}{r}} e^{-(nM/r)z} \tag{31}$$

The above clearly can be expanded in a power series in μ, which must converge for all μ for which

$$|Q| > 2,$$

and a solution of this inequality for μ leads to

$$|\mu| < r\frac{1-\sqrt{1-e^{-nMz/r}}}{1+\sqrt{1-e^{-nMz/r}}} \tag{32}$$

For $n = 1$, the preceding equation reduces to (24). Therefore (31) has a convergent expansion in μ, which can be shown to dominate (28), as in the previous case. The theorem therefore has been established.

These series methods are also of use in the case of linear differential equations involving a parameter.

Example. The differential equation

$$\begin{aligned} y'' + (\lambda - \mu \cos 2z)y &= 0, \\ y(0) &= 0, \\ y'(0) &= 1, \end{aligned} \tag{33}$$

is known as Mathieu's equation. Its solutions are known as Mathieu functions. Evidently these are functions of the two parameters λ and μ but we shall hold λ fixed and seek a series expansion in μ. To apply the preceding theorem directly, one should first convert the equation to a system, but one can also proceed by letting

$$y = \sum_0^\infty y_n \mu^n. \tag{34}$$

In the theorem it was assumed without loss of generality that the initial conditions vanished. In the present problem they do not vanish, but one can now impose the initial conditions

$$y_0(0) = 0, \qquad y_0'(0) = 1,$$
$$y_n(0) = 0, \qquad y_n'(0) = 0, \qquad n = 1, 2, \cdots.$$

By inserting (34) in (33) we obtain the following recursion formulas:

$$y_0'' + \lambda y_0 = 0,$$
$$y_n'' + \lambda y_n = (\cos 2z)\, y_{n-1}, \qquad n \geqslant 1.$$

From these and the initial conditions one can easily show that

$$y_0 = \frac{\sin \sqrt{\lambda}\, z}{\sqrt{\lambda}},$$

$$y_1 = \frac{\sin (\sqrt{\lambda} - 2)\, z}{8(\sqrt{\lambda} - 1)} - \frac{\sin (\sqrt{\lambda} + 2)\, z}{8(\sqrt{\lambda} + 1)} + \frac{\sin \sqrt{\lambda}\, z}{4(\lambda - 1)},$$

$$y_2 = \frac{\sin (\sqrt{\lambda} - 4)\, z}{128(\sqrt{\lambda} - 1)(\sqrt{\lambda} - 2)} + \frac{\sin (\sqrt{\lambda} - 2)\, z}{32(\lambda - 1)(\sqrt{\lambda} + 1)}$$
$$+ \frac{32\lambda^3 - 350\lambda^2 + 313\lambda - 4}{16\lambda(\lambda - 1)^2(\lambda - 4)} \sin \sqrt{\lambda}\, z - \frac{z \cos \sqrt{\lambda}\, z}{16 \sqrt{\lambda}\, (\lambda - 1)}$$
$$- \frac{\sin (\sqrt{\lambda} + 2)\, z}{32(\lambda - 1)(\sqrt{\lambda} - 1)} + \frac{\sin (\sqrt{\lambda} + 4)\, z}{128(\sqrt{\lambda} + 1)(\sqrt{\lambda} + 2)}.$$

Although no general formula for y_n can be determined, one can determine as many terms in the series as are convenient and obtain reasonable approximations to the solutions of (34). The above solutions fail if $\sqrt{\lambda}$ takes on integral values; in these cases one can still determine the solutions of (33) as limiting cases. For the case $\lambda = 1$, we obtain

$$y_0 = \sin z$$

$$y_1 = \lim_{\lambda \to 1} \frac{2 \sin \sqrt{\lambda}\, z + (\sqrt{\lambda} + 1) \sin (\sqrt{\lambda} - 2)\, z}{8(\lambda - 1)} - \frac{\sin (\sqrt{\lambda} + 2)\, z}{8(\sqrt{\lambda} + 1)}$$
$$= \frac{z \cos z}{16} - \frac{\sin z}{16} - \frac{\sin 3z}{16}, \text{ etc.}$$

6.3 STABILITY THEORY

In general it is difficult to obtain qualitative information in dealing with nonlinear differential equations. In many physical problems the independent variable z represents time and the dependent variables x_i indicate the state of a system. Often it is not necessary to determine the explicit solution to a problem, but it is important to be able to say something about the solution as time becomes infinite. In many physical problems one is motivated by the feeling that a small change in the conditions of the problem should result in a comparably small change in the solution.

For example, we consider the linear equation

$$X' = AX + F(z), \tag{1}$$

where A is a constant matrix. If X_1 and X_2 are two solutions of (1) corresponding to different initial conditions, and if we let

$$Y = X_1 - X_2,$$

then

$$Y' = AY. \tag{2}$$

From the methods of Chapter 2 it follows that all solutions of (2) must approach zero as z tends to infinity, provided all roots of the characteristic polynomial $|A - \lambda I|$ are such that the real part of $\lambda_i < 0$ for all i. This is clear if one realizes that the solution of (2) corresponding to a particular λ_i is given by

$$Y = e^{\lambda_i z}C, \tag{3}$$

where C is an arbitrary constant vector. If λ_i is a multiple eigenvalue, (3) may contain also polynomial terms, but

$$\lim_{z \to \infty} z^n e^{\lambda_i z} = 0,$$

provided

$$\text{Real part } \lambda_i < 0.$$

If some one eigenvalue is such that real part $\lambda_i > 0$, then the general solution of (2) must grow beyond all bounds as z approaches infinity. In the former case we shall say that (1) is stable; and in the latter case that it is unstable.

The word stability is a very tricky word. No one definition seems to be adequate for all purposes. For that reason several definitions will be given for different kinds of stability.

DEFINITION 1. The solutions of

$$X' = F(X, z)$$

will be said to be stable in the sense of Lyapunov if for every positive ϵ we can find a $\delta(\epsilon)$ such that for any two solutions X_1 and X_2, such that

$\| X_1(z_0) - X_2(z_0) \| < \delta(\epsilon)$ for some z_0, then $\| X_1(z) - X_2(z) \| < \epsilon$ for all $z_0 < z < \infty$.

DEFINITION 2. The solutions of

$$X' = F(X, z)$$

will be said to be asymptotically stable if one can find a δ such that

$$\| X_1(z_0) - X_2(z_0) \| < \delta \text{ implies } \lim_{z \to \infty} \| X_1(z) - X_2(z) \| = 0.$$

It is clear that all solutions of (1) are asymptotically stable if all eigenvalues of A have negative real parts.

In some instances the concept of stability has to be related to the set of initial conditions. For example, the equation

$$x' = -x(1 - x)(2 - x)$$

has the following set of solutions:

$$x_0 = 0,$$
$$x_1 = 1,$$
$$x_2 = 2.$$

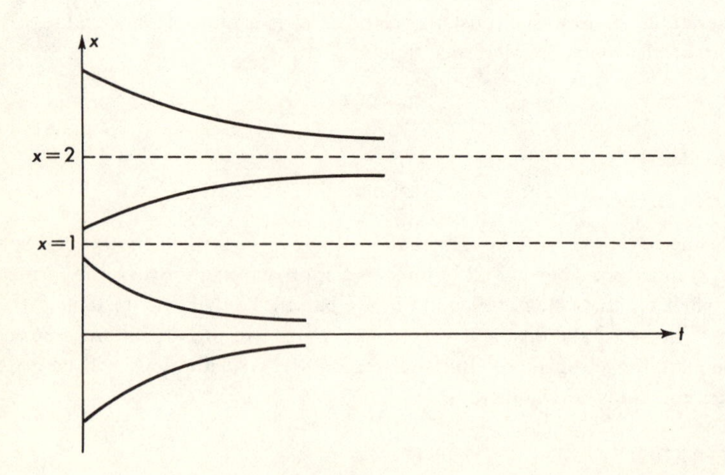

Fig. 6.1.

The preceding equation can be integrated explicitly, and one finds

$$(x - 1)^2 = \frac{1}{1 + c^2 e^{-2t}},$$

where c^2 is a constant of integration. One can now study the stability of the above three solutions. The results are most easily displayed graphically, as in Fig. 6.1.

Fig. 6.1 shows that x_0 and x_2 are stable solutions, whereas x_1 is unstable. Any solution with initial value greater than 1 approaches x_2. If the initial value is less than 1, the solution approaches x_0. If the initial value is precisely 1, the solution x_1 is obtained.

Accordingly, one should introduce an alternate set of definitions for stability.

DEFINITION 1′. The solution $X_1(z)$ of

$$X' = F(X, z)$$

will be said to be stable in the sense of Lyapunov with respect to a set of initial conditions A if for every positive ϵ we can find a $\delta(\epsilon)$ such that for any two solutions X_1 and X_2, with initial conditions in A, and $\| X_1(z_0) - X_2(z_0) \| < \delta(\epsilon)$ for some z_0, then $\| X_1(z) - X_2(z) \| < \epsilon$ for all $z_0 < z < \infty$.

A comparable definition can be furnished for asymptotic stability. In the previous example the solution x_0 is asymptotically stable with respect to the set of initial conditions $x(0) < 1$; x_2 is asymptotically stable with respect to the set of initial conditions $x(0) > 1$. In the sequel we shall often refer to a particular solution's being stable. This will generally mean stable with respect to some set of initial conditions sufficiently near the given solution.

The solutions of the system

$$X' = \begin{pmatrix} 0 & 1 \\ -1 & 0 \end{pmatrix} X$$

are no longer asymptotically stable, since the eigenvalues of the matrix are $\pm i$. However, the solutions are stable in the sense of Lyapunov. To show that, we must merely show that a solution which is small initially remains small. For a linear homogeneous equation this is equivalent to Lyapunov stability. Suppose $X(0)$ is given by

$$X(0) = \begin{pmatrix} c_1 \\ c_2 \end{pmatrix}.$$

Then

$$X = \begin{pmatrix} c_2 \sin z + c_1 \cos z \\ c_2 \cos z - c_1 \sin z \end{pmatrix}$$

and

$$\| X \| \leqslant 2 \mid c_1 \mid + 2 \mid c_2 \mid,$$
$$\| X(0) \| = \mid c_1 \mid + \mid c_2 \mid.$$

Therefore, in order for $\| X \| < \epsilon$, we require that $\delta(\epsilon) = \epsilon/2$.

This type of problem arises very often in the study of automatic control systems. In the case of linear equations with constant coefficients, the problem can be reduced to a study of the roots of the equation

$$| A - \lambda I | = 0.$$

But finding the roots of a higher order algebraic equation can be very tedious and timeconsuming. A number of techniques have been developed for deciding whether or not all roots have negative real parts, without finding the roots explicitly. The most common of these is known as the Routh-Hurwitz criterion. We shall not discuss it here because it would carry us too far a field.

In dealing with linear systems, where the matrix depends on z, or even with nonlinear systems, matters can become much more difficult. However, under suitable restrictions much can be said. The following theorems will consider some situations.

THEOREM: The solutions of the system

$$\begin{aligned} X' &= AX + F(X, z), \\ X(0) &= X_0, \end{aligned} \tag{5}$$

are asymptotically stable, provided all eigenvalues of A have negative real parts and $F(X, z)$ is a continuous function of X and z. Furthermore, $F(X, z)$ must be such that

$$|| F(X, z) || \leqslant k || X ||,$$

where k is sufficiently small and may depend on A.

From the assumptions on A it is clear that there must exist positive constants c and α such that

$$|| e^{Az} || \leqslant ce^{-\alpha z}.$$

We shall now use the method of variation of parameters to rewrite (5) as a Volterra integral equation. First let

$$X = e^{Az} Y,$$

so that (5) becomes

$$\begin{aligned} Y' &= e^{-Az} F(e^{Az} Y, z), \\ Y(0) &= X_0. \end{aligned}$$

By integration and multiplication by e^{Az} one obtains

$$X = e^{Az} X_0 + \int_0^z e^{A(z-t)} F(X, t)\, dt.$$

It follows that

$$\|X\| \leqslant \|e^{Az}\| \|X_0\| + \int_0^z \|e^{A(z-t)}\| \|F(X, t)\| \, dt$$

$$\leqslant c\|X_0\| e^{-\alpha z} + \int_0^z ce^{-\alpha(z-t)} k\|X\| \, dt, \tag{6}$$

which can be rewritten as

$$\frac{ke^{\alpha z}\|X\|}{\|X_0\| + k\int_0^z e^{\alpha t}\|X\| \, dt} \leqslant ck. \tag{7}$$

Now (7) can be integrated so that

$$ln \frac{\|X_0\| + k\int_0^z e^{\alpha t}\|X\| \, dt}{\|X_0\|} \leqslant ckz,$$

and one obtains

$$\|X_0\| + k\int_0^z e^{\alpha t}\|X\| \, dt \leqslant \|X_0\| e^{ckz}.$$

Combining this last inequality with (6) leads to

$$\|X\| \leqslant c\|X_0\| e^{(ck-\alpha)z}. \tag{8}$$

From (8) it is clear that if k is so small that $ck - \alpha < 0$, then

$$\lim_{z\to\infty} X = 0.$$

COROLLARY: The solutions of

$$X' = (A + B(z))X$$

and (9)

$$X(0) = X_0$$

are asymptotically stable if $B(z)$ is a continuous function of z and $\|B(z)\| \leqslant k$, where k is sufficiently small.

The proof of this corollary follows by an immediate application of the preceding theorem.

In some problems one is concerned not so much with questions of stability as with boundedness. The following two examples will demonstrate some of the subtleties that can be encountered.

The system

$$X' = \begin{pmatrix} 0 & 1 \\ -1 & -\frac{2}{z} \end{pmatrix} X \tag{10}$$

has the linearly independent solutions

$$X_1 = \begin{pmatrix} \dfrac{\sin z}{z} \\ \dfrac{z \cos z - \sin z}{z^2} \end{pmatrix}, \qquad X_2 = \begin{pmatrix} \dfrac{\cos z}{z} \\ -\dfrac{z \sin z + \cos z}{z^2} \end{pmatrix}.$$

Clearly, for large values of z, both solutions become arbitrarily small.

We now consider the very similar system

$$Y' = \begin{pmatrix} 0 & 1 \\ -1 & \dfrac{2}{z} \end{pmatrix} Y,$$

whose solutions are given by

$$Y_1 = \begin{pmatrix} z \sin z + \cos z \\ z \cos z \end{pmatrix}, \qquad Y_2 = \begin{pmatrix} z \cos z - \sin z \\ -z \sin z \end{pmatrix}.$$

Here, however, both solutions become large with growing z.

One might have been led to believe, on purely intuitive grounds, that for large values of z the solutions of these systems should resemble those of

$$X' = \begin{pmatrix} 0 & 1 \\ -1 & 0 \end{pmatrix} X, \tag{12}$$

whose solutions are

$$X_1 = \begin{pmatrix} \sin z \\ \cos z \end{pmatrix}, \qquad X_2 = \begin{pmatrix} \cos z \\ -\sin z \end{pmatrix}$$

and are bounded for all z. The following theorem helps to clarify this situation.

THEOREM: Let A be a constant matrix, all of whose eigenvalues have negative real parts, or even nonpositive real parts, provided the eigenvalues are simple and $B(z)$ is a matrix that is a differentiable function of z such that

$$\lim_{z \to \infty} B(z) = 0 \quad \text{and} \quad \int_0^\infty \| B' \| \, dt$$

exists. $C(z)$ will be taken to be a matrix is a continuous function of z, and

$$\int_0^\infty \| C \| \, dt$$

exists.

We shall also suppose that the eigenvalues of $A + B(z)$, which are, of course, functions of z, have negative real parts or even nonpositive real parts, provided the eigenvalues are simple.

Then the solutions of

$$X' = (A + B(z) + C(z))\,X$$

and

$$X(0) = X_0 \tag{13}$$

are stable in the sense of Lyapunov, or equivalently in this case, bounded.

The proof is very similar to the one for the preceding theorem. But first it will be convenient to diagonalize the matrix $A + B(z)$. We shall suppose that all of its eigenvalues $\lambda(z)$ are simple for all z. The theorem remains true in more general cases, but the proof must be modified. It must be possible to find a nonsingular matrix $T(z)$ such that

$$T^{-1}(z)\,(A + B(z))T(z) = D(z),$$

where $D(z)$ is diagonal. If we let

$$X = T(z)Y,$$

we find that

$$Y' = D(z)Y + E(z)Y, \tag{14}$$

where

$$E(z) = T^{-1}(z)C(z)T(z) - T^{-1}(z)T'(z).$$

From the fact that $\lim_{z\to\infty} B(z) = 0$, it follows that $D(z)$ approaches a diagonal matrix whose eigenvalues are equal to the eigenvalues of A, and $T(z)$ approaches a constant nonsingular matrix. Then

$$\int_0^\infty \| T^{-1}(z)\, C(z)\, T(z) \| \, dz < \infty$$

from the hypothesis on $C(z)$.

Since

$$\int_0^\infty \| B' \| \, dt < \infty,$$

it follows that

$$\int_0^\infty \| T^{-1}(z)\, T'(z) \| \, dz < \infty,$$

so that

$$\int_0^\infty \| E(z) \| \, dz < \infty.$$

We shall assign to the above integral the value M. From the hypotheses on the eigenvalues of $A + B(z)$ there must exist a constant c such that

$$\left\| \exp\left(\int_0^z D(t)\, dt\right) \right\| \leqslant c.$$

for all $z \geqslant 0$. Now (14) can be rewritten as a Volterra integral equation by first letting

$$Y = \exp\left(\int_0^z D(t)\,dt\right) W.$$

Then

$$W' = \exp\left(-\int_0^z D(t)\,dt\right) E(z) \exp\left(\int_0^z D(t)\,dt\right) W.$$

The above can be integrated, and one then finds for Y,

$$\begin{aligned} Y &= \exp\left(\int_0^z D(t)\,dt\right) Y_0 + \int_0^z \exp\left(\int_0^z D(t)\,dt\right) \exp\left(-\int_0^\tau D(t)\,dt\right) E(\tau)\, Y\, d\tau \\ &= \exp\left(\int_0^z D(t)\,dt\right) Y_0 + \int_0^z \exp\left(\int_\tau^z D(t)\,dt\right) E(\tau)\, Y\, d\tau. \end{aligned}$$

From this we find that

$$\|Y\| \leqslant c\,\|Y_0\| + \int_0^z c\,\|E(\tau)\|\,\|Y\|\,d\tau, \tag{15}$$

so that

$$\frac{\|E(z)\|\,\|Y\|}{\|Y_0\| + \int_0^z \|E(\tau)\|\,\|Y\|\,d\tau} \leqslant c\,\|E(z)\|.$$

By an integration it follows that

$$\|Y_0\| + \int_0^z \|E(\tau)\|\,\|Y\|\,d\tau \leqslant \|Y_0\| \exp\left(c\int_0^z \|E(t)\|\,dt\right) \leqslant \|Y_0\|\,e^{cM}.$$

From the preceding statement and (15) it follows that

$$\|Y\| \leqslant c\,\|Y_0\|\,e^{cM}, \tag{16}$$

and clearly, a similar statement applies to X, so that the theorem has been proved.

An application of the theorem to (10) and (11) shows that the solutions of (10) must be bounded and that the same could not be expected in the case of (11).

Another important stability concept is orbital stability. This arises in problems where, as the name suggests, closed orbits can be found; that is, the components $x_1(z)$, $x_2(z)$, $\cdots$ $x_n(z)$ of the solution vector of

$$X' = F(X, z)$$

trace out a trajectory in an n-dimensional $x_1, x_2, \cdots x_n$ space. If this trajectory yields a closed curve, it is called an "orbit." For example, the system

$$X' = \begin{pmatrix} 0 & 1 \\ -1 & 0 \end{pmatrix} X \tag{17}$$

has the solution

$$x_1(z) = c_1 \sin z + c_2 \cos z,$$
$$x_2(z) = c_1 \cos z - c_2 \sin z,$$

so that

$$x_1^2 + x_2^2 = (c_1^2 + c_2^2).$$

In this case the orbit is a circle. The radius of the circle depends on the initial conditions. The motion of the system is periodic with period 2π.

The following physical problem results in a system of type (17). Suppose a mass m is suspended from a spring, of spring constant k, and y represents the deflection from the rest position. The resultant differential equation of motion is

$$m \frac{d^2y}{dt^2} + ky = 0.$$

If we let

$$y = x_1,$$
$$y' = x_2,$$
$$z = \sqrt{\frac{k}{m}}\, t,$$

we obtain (17). It is clear that in this case we expect some sort of periodic motion. The period of this motion is given by 2π in the z variable and $\sqrt{m/k}\, 2\pi$ in the t variable. A unique feature of this problem is that the period of the motion is independent of the initial conditions.

The following problem is in some ways more typical of physical problems:

$$\frac{d^2y}{dz^2} + \operatorname{sgn} y = 0.$$

The function sgn y is defined as

$$\begin{aligned} \operatorname{sgn} y &= 1, & y &> 0, \\ &= 0, & y &= 0, \\ &= -1, & y &< 0. \end{aligned}$$

By letting

$$y = x_1,$$
$$y' = x_2,$$

we can rewrite the equation as

$$X' = \begin{pmatrix} x_2 \\ -\operatorname{sgn} x_1 \end{pmatrix}. \tag{18}$$

We now impose the following initial conditions:

$$\begin{aligned} x_1(0) &= 0, \\ x_2(0) &= a, \end{aligned} \tag{19}$$

where a is a positive constant. One can verify that

$$\begin{aligned} x_1 &= az - \tfrac{1}{2} z^2 \\ x_2 &= a - z, \end{aligned} \tag{20}$$

for

$$0 \leqslant z \leqslant 2\,a.$$

In this interval, $x_1 \geqslant 0$. Now

$$x_1(2a) = 0,$$
$$x_2(2a) = -a,$$

and one finds that

$$\begin{aligned} x_1 &= (z - 2a)\,(\tfrac{1}{2} z - 2a), \\ x_2 &= z - 3a, \end{aligned} \tag{21}$$

for

$$2a \leqslant z \leqslant 4a.$$

Clearly,

$$x_1(4a) = 0,$$
$$x_2(4a) = a,$$

and from (19) we see that

$$x_1(0) = x_1(4a),$$
$$x_2(0) = x_2(4a),$$

so that this solution of (18) is periodic with period $4a$. In this case it is clear that the period does depend on the initial conditions.

From (20) one finds, by eliminating z, that

$$x_1 = \tfrac{1}{2}(a^2 - x_2^2), \qquad x_1 > 0,$$

and similarly from (21),

$$x_1 = -\tfrac{1}{2}(a^2 - x_2^2), \qquad x_1 < 0.$$

Upon combining these, one obtains

$$|x_1| = \tfrac{1}{2}(a^2 - x_2^2). \tag{22}$$

When the latter is plotted in an x_1, x_2 coordinate system, one obtains a closed orbit as shown in Fig. 6.2.

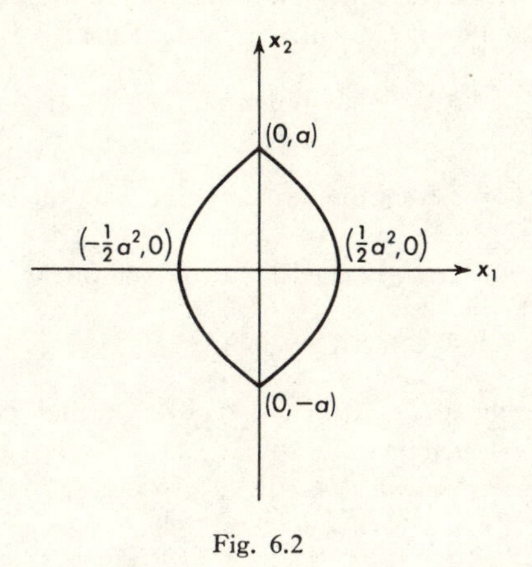

Fig. 6.2

The problems in (17) and (4) are identical, and it has been shown in the latter case that the solution is stable in the sense of Lyapunov. But this is entirely due to the fact that the period of the solution is independent of the initial conditions. A small change in the initial conditions will produce a small change in the solution and also in the orbit in the x_1, x_2 space. This conclusion is no longer valid for the solutions of (18). A small change in the initial value a may lead to radically different solutions ultimately. For example, for a particular initial value of a, we found the solution had period $4a$. If we now replace a by $a + (a/2n)$, where n is a positive integer, we find a solution of period $4a + (2a/n)$. We shall denote the solutions corresponding to the latter initial value by $\bar{x}_1$ and $\bar{x}_2$. For the latter initial condition we find at $z = 4an + 2a$,

$$\bar{x}_1(4an + 2a) = 0,$$

$$\bar{x}_2(4an + 2a) = a + \frac{a}{2n},$$

whereas for the same value of z, the former solution yields

$$x_1(4an + 2a) = x_1(2a) = 0,$$
$$x_2(4an + 2a) = x_2(2a) = -a,$$

since its period is $4a$. Therefore

$$|| X - \bar{X} || = 2a + \frac{a}{2n},$$

and this situation will be repeated infinitely often. Evidently the solution is no longer stable in the sense of Lyapunov. However, a small change in a will result in only a small change in the orbit defined by (22). This type of stability is commonly called orbital stability, for which rigorous definition will now be given. We shall denote the distance of a point X in the n-dimensional $x_1, x_2, \cdots, x_n$ space from a closed orbit C by $d(X, C)$ and define it by

$$d(X, C) = \underset{Y \text{ on } C}{\text{minimum}} \mid X - Y \mid;$$

that is, the distance of X from the nearest point Y on the orbit C.

DEFINITION: The orbit C, defined by a solution of the differential equation

$$X' = F(X, z),$$

will be called orbitally stable, if for every ϵ we can find a $\delta(\epsilon)$ such that if an initial point X_0 is selected,

$$d(X_0, C) < \delta(\epsilon);$$

then

$$d(X, C) < \epsilon,$$

for all $z > z_0$.

DEFINITION: The orbitally stable orbit C of the preceding definition will be called asymptotically orbitally stable if

$$\lim_{z \to \infty} d(X, C) = 0.$$

An example of an asymptotically orbitally stable system is given by

$$\begin{aligned} \frac{dx_1}{dz} &= x_2 + x_1 (1 - x_1^2 - x_2^2), \\ \frac{dx_2}{dz} &= -x_1 + x_2 (1 - x_1^2 - x_2^2). \end{aligned} \tag{23}$$

To analyze the system, it is most convenient to convert it to polar coordinates:

$$r = (x_1^2 + x_2^2)^{1/2},$$

$$\theta = \tan^{-1} \frac{x_2}{x_1}.$$

One then finds that

$$\begin{aligned} \frac{dr}{dz} &= r(1 - r^2), \\ \frac{d\theta}{dz} &= -1. \end{aligned} \tag{24}$$

Clearly, the orbit

$$r = 1$$

is a valid solution. To test it for stability, one can find r explicitly and then

$$\frac{r}{(1 - r^2)^{1/2}} = \frac{e^z}{c},$$

where c is a constant of integration. Solving for r, one finds that

$$r = \frac{1}{(1 + c^2 e^{-2z})^{1/2}},$$

and clearly,

$$\lim_{z \to \infty} r = 1.$$

The only difficulty that can arise is if $r = 0$ initially, since this also satisfies the equation, but one can easily show that this is an unstable solution.

The system (24) was easy to analyze because explicit integrations were possible. In some problems it may be simple to read off specific solutions, but explicit integrations may no longer be feasible. Nevertheless it may be important to test the solution for stability. For example, once it is recognized that (24) has a solution

$$r = 1,$$

one can let

$$r = 1 + y.$$

Then the first equation in (24) becomes

$$\frac{dy}{dz} = -2y - 3y^2 - y^3. \tag{25}$$

This equation has a solution $y = 0$, and for small y, can be approximated by

$$\frac{dy}{dz} = -2y.$$

This equation has asymptotically stable solutions. Therefore any solution of (25) with sufficiently small initial values must also be asymptotically stable, provided the higher terms are negligible. That this is so was proved for (5), and clearly (25) satisfies the hypothesis imposed on (5).

We shall now restrict ourselves to autonomous systems; that is, systems of the form

$$X' = F(X), \tag{26}$$

where the right side does not contain the independent variable explicitly. This type of equation arises most frequently in physical problems, since the physical components generally depend on the state of the system but not on the actual time. We shall suppose that (26) has the solution

$$X = 0.$$

To test the solution for stability, it will be convenient to approximate the right side of (26) by a linear approximation.

We recall that for a function of a single variable with a continuous first derivative, we have from the mean value theorem:

$$f(x + h) = f(x) + hf'(x) + h\epsilon(h),$$

where

$$\lim_{h\to 0} \epsilon(h) = 0.$$

Similarly, for a function of several variables with continuous first partial derivatives, we have

$$\begin{aligned} f(x_1 + h_1, x_2 + h_2, \cdots, x_n + h_n) = f(x_1, x_2, \cdots, x_n) \\ + \sum_{i=1}^{n} h_i \frac{\partial f}{\partial x_i}(x_1, x_2, \cdots, x_n) \\ + \sum_{i=1}^{n} h_i\epsilon_i(h_1, h_2, \cdots, h_n) \end{aligned} \tag{27}$$

Where:

$$\lim_{\|h\|\to 0} \epsilon_i(h_1, h_2, \cdots, h_n) = 0,$$

$$\| h \| = \sum_{i=1}^{n} | h_i |.$$

If the components of $F(X)$ in (26) are functions with continuous first partial derivatives, then (26) can be rewritten in the form

$$X' = F_x(0)X + E(X)X. \tag{28}$$

Here $F_x(0)$ is a matrix defined by

$$F_x(0) = \left(\frac{\partial f_i(0, 0, \cdots, 0)}{\partial x_j}\right),$$

where the f_i are the components of $F(X)$ and $E(X)$ is a matrix such that

$$\lim_{\|X\| \to 0} \| E(X) \| = 0.$$

This follows immediately by applying (27) to each of the components of $F(X)$; (28) is now in the form (5). We first suppose that the solutions of

$$X' = F_x(0)X$$

are asymptotically stable. Then, if $\| X_0 \|$ is sufficiently small, it will follow that

$$\| E(X)X \| \leqslant k \| X \|$$

for a sufficiently small k, so that by the theorem the solutions of (28) are also asymptotically stable. This leads us to a new theorem.

THEOREM: The solution $X = 0$ of the equation

$$X' = F(X)$$

(provided, of course, that this is a solution) is asymptotically stable for all sufficiently small initial values if $F(X)$ has continuous first derivatives and if the (so-called) equation of first variation,

$$X' = F_x(0)X,$$

has asymptotically stable solutions.

EXAMPLE. Test for stability the solution $X = 0$ of the system

$$X' = \begin{pmatrix} x_2 \\ -cx_2 - \sin x_1 \end{pmatrix}. \tag{29}$$

Now (29) can be rewritten as

$$X' = \begin{pmatrix} 0 & 1 \\ -1 & -c \end{pmatrix} X + \begin{pmatrix} 0 & 0 \\ \dfrac{x_1 - \sin x_1}{x_1} & 0 \end{pmatrix} X.$$

In this case we have

$$\| E(X) \| = \left\| \begin{pmatrix} 0 & 0 \\ \dfrac{x_1 - \sin x_1}{x_1} & 0 \end{pmatrix} \right\| = \left| \frac{x_1 - \sin x_1}{x_1} \right|$$

and

$$\lim_{\|X\| \to 0} \| E(X) \| = 0.$$

The linear system

$$X' = \begin{pmatrix} 0 & 1 \\ -1 & -c \end{pmatrix} X$$

has the eigenvalues

$$\lambda = \frac{-c \pm \sqrt{c^2 - 4}}{2}$$

and has asymptotically stable solutions for all $c > 0$. Therefore a solution of (29) for which $\| X_0 \|$ is sufficiently small is asymptotically stable.

6.4 LYAPUNOV'S DIRECT METHOD

A method for stability analysis that has come into prominence in the last few years is Lyapunov's direct method, or as it is sometimes called, Lyapunov's second method. It can be interpreted in purely physical terms. The total energy of a physical system at a point of stable equilibrium must be at a minimum. For example, the total kinetic energy of a particle of mass m and velocity x' is $\frac{1}{2} m(x')^2$. The total stored energy of a spring, of spring constant k and extension x, is $\frac{1}{2} kx^2$. Therefore, if a mass m is suspended from a spring and allowed to vibrate, the total energy of the system is given by

$$V = \tfrac{1}{2} m(x')^2 + \tfrac{1}{2} kx^2.$$

The differential equation of motion is given by

$$mx'' + cx' + kx = 0,$$

where the term cx' represents the resistance of motion due to friction and similar restraining forces. One now computes the derivative of V and finds that

$$\begin{aligned} V' &= mx'x'' + kxx' = x'(mx'' + kx) \\ &= x'(-cx') = -c(x')^2, \end{aligned}$$

by use of the differential equation. Since on physical grounds c is a positive constant, V' is a negative quantity, so that V is decreasing with time. But clearly V can never become negative. A decreasing function with a lower bound must have a greatest lower bound. The total energy V has the greatest lower bound 0, which is attained when the system reaches its stable rest position $x = 0$. From this it follows, as will be shown shortly, that the solution $x = 0$ is asymptotically stable.

The essence of the method of Lyapunov is to find a function $V(x_1, x_2, \cdots, x, z)$ that behaves like an energy function. But first the terms to be used have to be made more precise.

DEFINITION: A continuous function $V(x_1, x_2, \cdots, x_n, z)$ will be called positive definite if

$$\lim_{\|X\| \to 0} V(x_1, x_2, \cdots, x_n, z) = 0$$

and there exists a function $\phi(|X|)$ such that

$$V(x_1, x_2, \cdots, x_n, z) \geqslant \phi(|X|).$$

The function $\phi(|X|)$ must be a monotonically increasing function of its argument $|X|$ and $\phi(0) = 0$.

A function $V(x_1, x_2, \cdots, x_n, z)$ will be called *negative definite* if there exists a function $\phi(|X|)$ of the type described above such that

$$V(x_1, x_2, \cdots, x_n, z) \leqslant -\phi(|X|).$$

THEOREM: Consider

$$X' = F(X, z),$$

where $F(X, z)$ is a continuous function of its arguments and

$$F(0, z) = 0.$$

The solution $X = 0$ will be stable in the sense of Lyapunov if a positive definite function $V(x_1, x_2, \cdots, x_n, z)$ can be found such that

$$V' = \frac{\partial V}{\partial z} + \sum_{i=1}^{n} \frac{\partial V}{\partial x_i} f_i(x_1, x_2, \cdots, x_n, z) \leqslant 0,$$

where the f_i are the components of $F(X, z)$ for all trajectories that are initially sufficiently close to the origin.

We can certainly select an ϵ and an $|X_0| < \epsilon$ such that we also have

$$V(x_1(0), x_2(0), \cdots, x_n(0), 0) < \phi(\epsilon).$$

We now suppose that for some value z, the solution X with initial value X_0 will have $|X| = \epsilon$. Then it follows that at the value of z,

$$V(x_1, x_2, \cdots, x_n, z) \geqslant \phi(\epsilon).$$

But by hypothesis, V is a decreasing function of z, so that if initially

$$V < \phi(\epsilon),$$

it must remain so. Hence the fact that $|X| = \epsilon$ is contradicted, and so

$$|X| < \epsilon, \qquad \text{for all } z \geqslant 0.$$

This implies that $X = 0$ is stable in the sense of Lyapunov.

There is a companion theorem that relates to asymptotic stability.

THEOREM: The solution $X = 0$ under the hypotheses of the preceding theorem will be asymptotically stable if V' is not merely nonpositive but actually negative definite.

It follows immediately from the preceding theorem that $X = 0$ is stable in the sense of Lyapunov. By hypothesis we can find functions $\phi(|X|)$ and $\psi(|X|)$ such that

$$V \geqslant \phi(|X|), \qquad V' \leqslant -\psi(|X|).$$

Since V is a decreasing function, it must follow that $\lim_{z\to\infty} V$ exists and is, say, $l \geqslant 0$. If $l = 0$, the asymptotic stability has been proved. Otherwise there must exist a positive quantity ρ such that

$$|X| \geqslant \rho > 0$$

and

$$\begin{aligned} V(x_1, x_2, \cdots, x_n, z) &\geqslant \phi(\rho) > 0, \\ V'(x_1, x_2, \cdots, x_n, z) &\leqslant -\psi(\rho). \end{aligned} \tag{1}$$

By integration it follows from the second of (1) that

$$V(x_1, x_2, \cdots, x_n, z) \leqslant V(x_1(0), x_2(0), \cdots, x_n(0), 0) - z\psi(\rho). \tag{2}$$

Since for $\rho > 0$, $\psi(\rho) \neq 0$, it follows from (2) that for some z,

$$V < 0,$$

contradicting (1). Therefore $l = 0$ and the solution is asymptotically stable.

The real weakness of Lyapunov's direct method is that no general technique is known for the construction of the functions $V(x_1, x_2, \cdots, x_n, z)$, which are known as Lyapunov functions.

EXAMPLE 1. We shall again consider the problem solved at the end of the preceding section. The system was

$$X' = \begin{pmatrix} x_2 \\ -cx_2 - \sin x_1 \end{pmatrix}. \tag{3}$$

A suitable Lyapunov function is given by

$$V = \frac{\left(\sqrt{c}\, x_1 + \dfrac{1}{\sqrt{c}} x_2\right)^2}{2} + \frac{x_1^2 + \frac{1}{2} x_2^2}{c}.$$

One can show that a function $\phi(|X|)$ exists such that

$$V \geqslant \phi\ (|X|).$$

By using (3), one finds

$$V' = -(x_1^2 + x_2^2) - \left(x_1 + \frac{2}{c} x_2\right)(\sin x_1 - x_1).$$

It is easy to show that this function is negative definite. It follows, therefore, that the solution $X = 0$ is asymptotically stable. The mystery of how the preceding Lyapunov function was constructed will now be cleared up.

We shall now discuss a general method for finding Lyapunov functions for asymptotically stable linear systems with constant coefficients. This will show that for stable linear systems with constant coefficients, Lyapunov functions do exist. This method also applies to almost linear systems like (3). In Section 1.6 it was shown that every matrix with distinct eigenvalues can be diagonalized; that is, a nonsingular matrix T can be found such that

$$T^{-1}AT = D.$$

But if A has complex eigenvalues, T must be a complex matrix. In most applications to differential equations A is real, so that its eigenvalues are real or occur in complex conjugate pairs. But it may not be convenient to operate with complex elements. In such cases one can find a slightly different normal form, operating only with real terms.

Now we suppose that A is a real matrix with distinct eigenvalues, and we denote its real eigenvalues by $\lambda_1, \lambda_2, \cdots, \lambda_k$ and its complex eigenvalues by $u_1, u_2, \cdots, u_m$. We suppose that they are so indexed that

$$u_1 = \bar{u}_2,$$
$$u_3 = \bar{u}_4,$$
$$u_{m-1} = \bar{u}_m,$$

and define a_i and b_i as follows:

$$u_1 = a_1 + ib_1,$$
$$u_2 = a_1 - ib_1, \qquad \text{etc.}$$

Corresponding to each of the real eigenvalues we can find real eigenvectors L_j such that

$$AL_j = \lambda_j L_j.$$

Corresponding to the complex eigenvalues we find complex eigenvectors $\mathscr{L}_j$ such that, for example,

$$A\mathscr{L}_1 = u_1\mathscr{L}_1.$$

If we break up the $\mathscr{L}_j$ into their real and imaginary components, we obtain, for example,

$$\mathscr{L}_1 = R_1 + iI_1,$$

so that

$$A\mathscr{L}_1 = AR_1 + iAI_1 = (a_1 + ib_1)(R_1 + iI_1)$$
$$= (a_1R_1 - b_1I_1) + i(b_1R_1 + a_1I_1).$$

By equating real and imaginary terms, we have

$$AR_1 = a_1R_1 - b_1I_1,$$
$$AI_1 = b_1R_1 + a_1I_1,$$

and similarly for all other complex eigenvalues. We now construct a matrix T whose columns are all real and are formed with all column vectors L_i, R_i, I_i. T is also nonsingular. The matrix $T^{-1}AT$ is no longer diagonal, but is almost so. It has the form

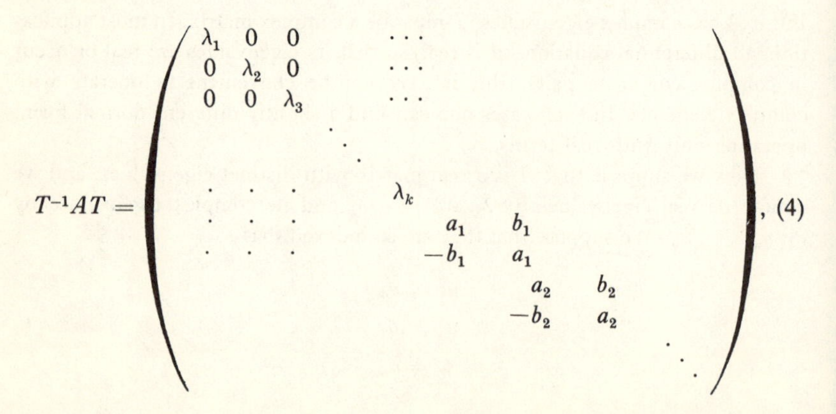

$$T^{-1}AT = \begin{pmatrix} \lambda_1 & 0 & 0 & & \cdots & & & & \\ 0 & \lambda_2 & 0 & & \cdots & & & & \\ 0 & 0 & \lambda_3 & & \cdots & & & & \\ & & & \ddots & & & & & \\ \cdot & \cdot & \cdot & & \lambda_k & & & & \\ \cdot & \cdot & \cdot & & & a_1 & b_1 & & \\ \cdot & \cdot & \cdot & & & -b_1 & a_1 & & \\ & & & & & & & a_2 & b_2 \\ & & & & & & & -b_2 & a_2 \\ & & & & & & & & \ddots \end{pmatrix}, \qquad (4)$$

where, corresponding to each real eigenvalue, diagonal elements λ_i enter; and corresponding to each complex eigenvalue, 2×2 blocks occur of the form

$$\begin{matrix} a_i & -b_i \\ b_i & a_i \end{matrix}$$

and all other elements vanish.

EXAMPLE 2.

$$A = \begin{pmatrix} -1 & 2 & 3 \\ 0 & -2 & -4 \\ 0 & 1 & -2 \end{pmatrix},$$

$$| A - \lambda I | = -(1 + \lambda)(\lambda^2 + 4\lambda + 8),$$

$$\lambda_1 = -1,$$
$$u_1 = -2 + 2i,$$
$$u_2 = -2 - 2i,$$

$$L_1 = \begin{pmatrix} 1 \\ 0 \\ 0 \end{pmatrix}, \quad \mathscr{L}_1 = \begin{pmatrix} 1 - 2i \\ 2i \\ 1 \end{pmatrix}, \quad \mathscr{L}_2 = \begin{pmatrix} 1 + 2i \\ -2i \\ 1 \end{pmatrix}.$$

We now find that

$$R_1 = \begin{pmatrix} 1 \\ 0 \\ 1 \end{pmatrix}, \quad I_1 = \begin{pmatrix} -2 \\ 2 \\ 0 \end{pmatrix},$$

so that

$$T = \begin{pmatrix} 1 & 1 & -2 \\ 0 & 0 & 2 \\ 0 & 1 & 0 \end{pmatrix}, \quad T^{-1} = \begin{pmatrix} 1 & 1 & -1 \\ 0 & 0 & 1 \\ 0 & \frac{1}{2} & 0 \end{pmatrix}.$$

A calculation shows that

$$T^{-1}AT = \begin{pmatrix} -1 & 0 & 0 \\ 0 & -2 & 2 \\ 0 & -2 & -2 \end{pmatrix}.$$

We shall now return to the differential equation

$$X' = AX \tag{5}$$

which will by hypothesis be assumed to be asymptotically stable. We shall first find a real nonsingular matrix T such that $T^{-1}AT$ is in the normal form (4). Then

$$X = TY,$$

so that

$$Y' = T^{-1}ATY = DY, \tag{6}$$

where D is of form (4).. For stability we require that all diagonal elements of D be negative.

Now we shall seek a Lyapunov function in the form

$$V = (Y, BY), \tag{7}$$

using inner product notation. B will be a suitable matrix, which we shall suppose to be symmetric.

$$\begin{aligned} V' &= (Y', BY) + (Y, BY') \\ &= (DY, BY) + (Y, BDY) \\ &= (Y, (D^TB + BD)Y). \end{aligned} \tag{8}$$

(D^T denotes the transpose of D.)

In order to be sure that V' is negative definite, we shall require that

$$V' = -(Y, Y) = -\sum_1^n y_i^2, \tag{9}$$

where the y_i are the components of Y. In order for (8) and (9) to agree, we now impose the condition that

$$D^TB + BD = -I,$$

where I is the identity. The last is an equation for B. One can easily see, and verify, that a solution for B is given by a diagonal matrix:

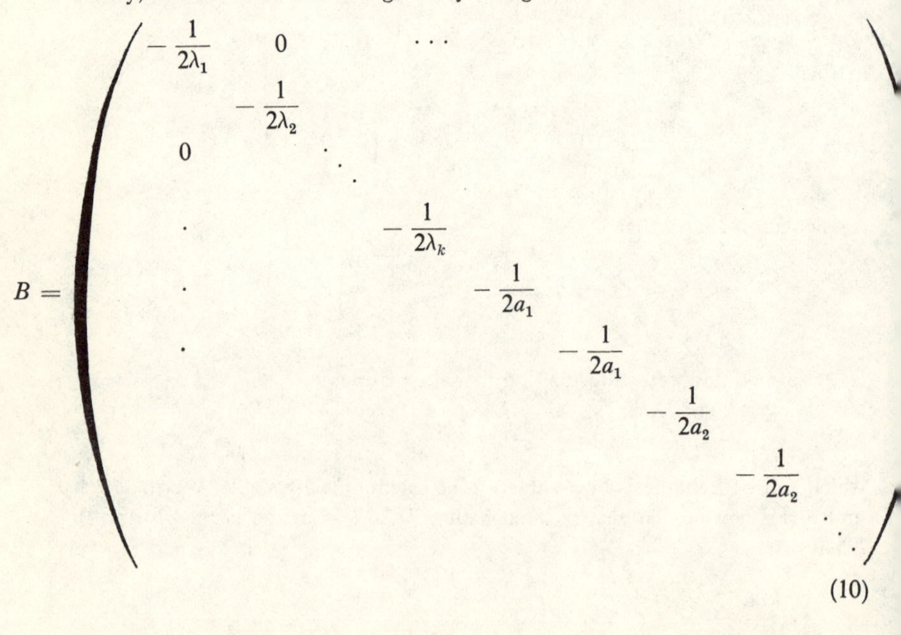

$$B = \begin{pmatrix} -\frac{1}{2\lambda_1} & 0 & \cdots & & & & & \\ & -\frac{1}{2\lambda_2} & & & & & & \\ 0 & & \ddots & & & & & \\ \cdot & & & -\frac{1}{2\lambda_k} & & & & \\ \cdot & & & & -\frac{1}{2a_1} & & & \\ \cdot & & & & & -\frac{1}{2a_1} & & \\ & & & & & & -\frac{1}{2a_2} & \\ & & & & & & & -\frac{1}{2a_2} \\ & & & & & & & & \ddots \end{pmatrix} \tag{10}$$

Since by hypothesis all λ_i and a_i are negative, all diagonal elements in B are positive, and therefore (7) consists of sums of positive terms so that V is positive definite:

$$V = -\frac{1}{2\lambda_1}y_1^2 - \frac{1}{2\lambda_2}y_2^2 - \cdots - \frac{1}{2\lambda_k}y_k^2 - \frac{1}{2a_1}(y_{k+1}^2 + y_{k+2}^2) + \cdots.$$

EXAMPLE 3. Construct a Lyapunov function for the system

$$X' = \begin{pmatrix} -1 & 2 & 3 \\ 0 & -2 & -4 \\ 0 & 1 & -2 \end{pmatrix} X.$$

In Example 2 it was shown that if

$$T = \begin{pmatrix} 1 & 1 & -2 \\ 0 & 0 & 2 \\ 0 & 1 & 0 \end{pmatrix},$$

then

$$T^{-1}AT = \begin{pmatrix} -1 & 0 & 0 \\ 0 & -2 & 2 \\ 0 & -2 & -2 \end{pmatrix}.$$

We let

$$X = TY$$

and find

$$Y' = \begin{pmatrix} -1 & 0 & 0 \\ 0 & -2 & 2 \\ 0 & -2 & -2 \end{pmatrix} Y. \tag{11}$$

To find a Lyapunov function, we seek a matrix B such that

$$D^TB + BD = -I,$$

where D is defined in (11). A simple check shows that

$$B = \begin{pmatrix} \frac{1}{2} & 0 & 0 \\ 0 & \frac{1}{4} & 0 \\ 0 & 0 & \frac{1}{4} \end{pmatrix}.$$

The Lyapunov function is given by

$$V = (Y, BY) = \tfrac{1}{2}y_1^2 + \tfrac{1}{4}y_2^2 + \tfrac{1}{4}y_3^2$$

and

$$V' = -y_1^2 - y_2^2 - y_3^2.$$

By transforming back to the X variable, one finds a Lyapunov function for the original equation.

This result can be used to provide a different proof for a theorem proved in the preceding section. There it was shown that if the solutions of

$$X' = AX$$

are asymptotically stable, the same is true for the solutions of

$$X' = AX + F(X, z),$$

where

$$\| F(X, z) \| \leqslant k \| X \|$$

and k is sufficiently small.

A suitable Lyapunov function is given by

$$V = (X, BX) = X^T BX,$$

where B is so constructed that

$$A^T B + BA = -I.$$

V is certainly positive definite.

$$\begin{aligned} V' &= (X^T)BX + X^T BX' \\ &= (X^T A^T + F^T)BX + X^T B(AX + F) \\ &= X^T(A^T B + BA)X + F^T BX + X^T BF \\ &= -X^T X + F^T BX + X^T BF. \end{aligned}$$

Now

$$\| F^T BX + X^T BF \| \leqslant 2 \| F \| \ \| B \| \ \| X \| \leqslant 2k \| B \| \ \| X \|^2,$$

and for sufficiently small k,

$$2k \| B \| \ \| X \|^2 \leqslant \| X \|^2.$$

Therefore, for $k < \dfrac{1}{2 \| B \|}$, V' will be negative definite, so that the solution $X = 0$ is asymptotically stable.

Once a suitable Lyapunov function has been found, the stability of the solution is assured. But if no such function can be found, nothing is known. It is therefore of the utmost importance to develop some instability tests. One such is due to Chetayev.

THEOREM: Suppose one can find a differentiable function $V(x_1, x_2, \cdots, x_n)$, which in every sufficiently small neighborhood of the origin can take on positive as well as negative values, and such that

$$V' = \sum_{i=1}^{n} \frac{\partial V}{\partial x_i} f_i(x_1, x_2, \cdots, x_n) > 0,$$

where the f_i are the components of the right side of

$$X' = F(X), \qquad F(0) = 0.$$

(We suppose here that the differential equation is autonomous, that is F is independent of z.)

Suppose furthermore that in every neighborhood where $V \geqslant \alpha > 0$, $V' \geqslant \beta > 0$; then the solution $X = 0$ is unstable.

We consider a region where $V > 0$ and on whose boundary $V = 0$. For any initial point $X_0 \neq 0$ in that region, we must have $V(x_1(0), x_2(0), \cdots, x_n(0)) \geqslant \alpha > 0$ for some α. Therefore $V' \geqslant \beta > 0$ for some β. Since V is an increasing function, the point X cannot approach the boundary of the region and therefore certainly cannot approach the origin. Therefore $X = 0$ is an unstable solution.

EXAMPLE 4. The differential equation

$$X' = \begin{pmatrix} -1 & 0 \\ 0 & 1 \end{pmatrix} X + \begin{pmatrix} x_2^3 \\ x_1^2 \end{pmatrix}$$

clearly has the solution $X = 0$. The function

$$V = x_2^2 - x_1^2$$

takes on positive as well as negative values near $X = 0$.

$$V' = 2(x_1^2 + x_2^2) - 2x_1x_2(x_2^2 - x_1),$$

and clearly for small $|X|$, V' is positive definite, since the first term on the right is quadratic and the second is cubic. Therefore the solution $X = 0$ is unstable.

6.5 DIFFERENTIAL EQUATIONS WITH PERIODIC SOLUTIONS

In Chapter 5 the structure of linear differential equations with periodic coefficients was analyzed. The systems considered were of the form

$$\begin{aligned} X' &= A(z)\,X, \\ A(z + T) &= A(z), \end{aligned} \tag{1}$$

where T is the period of the coefficients. It was shown that if $\Phi(z)$ was a fundamental matrix solution of (1), and if μ was a root of the equation

$$|\Phi(T) - \mu\Phi(0)| = 0, \tag{2}$$

then one could find a solution of the form

$$X = e^{cz}P(z), \tag{3}$$

where c is defined by

$$e^{cT} = \mu,$$

and

$$P(z + T) = P(z).$$

A necessary and sufficient condition for (1) having a periodic solution of period T is that for at least one of the roots of (2), $\mu = 1$.

One can also obtain a criterion for asymptotic stability. All solutions of (1) will be asymptotically stable if and only if for all roots μ_i of (2), we have $|\mu_i| < 1$.

A similar analysis for the nonlinear equation

$$X' = F(X, z),$$

where

$$F(X, z + T) = F(X, z),$$

is in general impossible. But this type of problem does come up in many applications, in particular in celestial mechanics. It was Poincaré who first conceived the idea that a study of these equations can be greatly simplified if a more general class of problems is considered. One tries to embed a specific problem in a broad class of problems of the type

$$X' = F(X, z, \mu), \tag{4}$$

where μ is a parameter. It may be clear that for $\mu = 0$, (4) does have a periodic solution. If it can then be shown that for some neighborhood of $\mu = 0$, periodic solutions can still be obtained, one can build up a family of periodic solutions depending on μ. Then we hope that if for, say, $\mu = 1$, (4) reduces to the case of interest, we can extend the family of periodic solutions up to $\mu = 1$. This process may be compared to the theorem relating to (15) of Section 6.2. The case treated there is simpler because no periodicity requirement is to be invoked.

Before stating and proving a theorem it will be desirable to demonstrate the ideas on the basis of a concrete example.

EXAMPLE. Investigate the periodic solutions of period 2π of

$$u'' + \omega^2(u + \mu u^3) = \sin z,$$

where μ is small and ω is not an integer. By letting

$$u = x_1,$$
$$u' = x_2,$$

we obtain the system

$$X' = \begin{pmatrix} 0 & 1 \\ -\omega^2 & 0 \end{pmatrix} X + \begin{pmatrix} 0 \\ \sin z - \mu\omega^2 x_1^3 \end{pmatrix}. \tag{5}$$

For $\mu = 0$, we obtain as a general solution:

$$X(z) = \begin{pmatrix} \dfrac{\sin z}{\omega^2 - 1} \\ \dfrac{\cos z}{\omega^2 - 1} \end{pmatrix} + \begin{pmatrix} c_1 \sin \omega z + c_2 \cos \omega z \\ \omega c_1 \cos \omega z - \omega c_2 \sin \omega z \end{pmatrix}.$$

In order for this solution to have period 2π, it is necessary to select the initial conditions

$$X(0) = \begin{pmatrix} 0 \\ \dfrac{1}{\omega^2 - 1} \end{pmatrix},$$

so that $c_1 = c_2 = 0$ and the solution of period 2π is given by

$$X(z) = \begin{pmatrix} \dfrac{\sin z}{\omega^2 - 1} \\ \dfrac{\cos z}{\omega^2 - 1} \end{pmatrix}. \tag{6}$$

We shall now seek a periodic solution of (5), which for small values of μ will be close to (6). To ensure the periodicity, however, the initial values will have to be modified suitably. We therefore seek a periodic solution

$$X(z, \alpha, \mu),$$

where $X(z, 0, 0)$ is given by (6) such that

$$X(2\pi, \alpha, \mu) = X(0, \alpha, \mu) = X(0, 0, 0) + \alpha.$$

Therefore α represents the necessary change in the initial conditions to ensure the periodicity, to be determined from the implicit equation

$$X(2\pi, \alpha, \mu) - X(0, 0, 0) - \alpha = 0. \tag{7}$$

For $\mu = 0$, we see that $\alpha = 0$ is a solution. To obtain a clear view of (7), we shall write out the components in detail:

$$\begin{pmatrix} x_1(2\pi, \alpha_1, \alpha_2, \mu) \\ x_2(2\pi, \alpha_1, \alpha_2, \mu) \end{pmatrix} - \begin{pmatrix} 0 \\ \dfrac{1}{\omega^2 - 1} \end{pmatrix} - \begin{pmatrix} \alpha_1 \\ \alpha_2 \end{pmatrix} = 0.$$

This leads to the two simultaneous equations

$$x_1(2\pi, \alpha_1, \alpha_2, \mu) - \alpha_1 = 0,$$
$$x_2(2\pi, \alpha_1, \alpha_2, \mu) - \frac{1}{\omega^2 - 1} - \alpha_2 = 0,$$

for α_1 and α_2. From the structure of (5) these must be analytic functions of α_1, α_2, μ, and we can expand in terms of the small quantities α_1, α_2, μ :

$$x_1(2\pi, 0, 0, 0) + \frac{\partial x_1}{\partial \alpha_1}(2\pi, 0, 0, 0)\,\alpha_1 + \frac{\partial x_1}{\partial \alpha_2}(2\pi, 0, 0, 0)\,\alpha_2 + \frac{\partial x_1}{\partial \mu}(2\pi, 0, 0, 0)\,\mu - \alpha_1 + \cdots = 0,$$

$$x_2(2\pi, 0, 0, 0) + \frac{\partial x_2}{\partial \alpha_1}(2\pi, 0, 0, 0)\,\alpha_1 + \frac{\partial x_2}{\partial \alpha_2}(2\pi, 0, 0, 0)\,\alpha_2 + \frac{\partial x_2}{\partial \mu}(2\pi, 0, 0, 0)\,\mu - \frac{1}{\omega^2 - 1} - \alpha_2 + \cdots = 0.$$

The terms indicated by $+ \cdots$ will be of a higher order of magnitude for small α_1, α_2, and μ, and can by the use of suitable remainder terms be shown to be negligible. We now use the fact that

$$x_1(2\pi, 0, 0, 0) = 0,$$
$$x_2(2\pi, 0, 0, 0) = \frac{1}{\omega^2 - 1},$$

so that the above system can be rewritten in the form

$$\begin{pmatrix} \dfrac{\partial x_1}{\partial \alpha_1}(2\pi, 0, 0, 0) - 1 & \dfrac{\partial x_1}{\partial \alpha_2}(2\pi, 0, 0, 0) \\ \dfrac{\partial x_2}{\partial \alpha_1}(2\pi, 0, 0, 0) & \dfrac{\partial x_2}{\partial \alpha_2}(2\pi, 0, 0, 0) - 1 \end{pmatrix} \begin{pmatrix} \alpha_1 \\ \alpha_2 \end{pmatrix} = \begin{pmatrix} \dfrac{\partial x_1}{\partial \mu}(2\pi, 0, 0, 0) \\ \dfrac{\partial x_2}{\partial \mu}(2\pi, 0, 0, 0) \end{pmatrix} \mu + \cdots. \tag{8}$$

The system (8) can be expected to have a solution, provided its determinant does not vanish; that is ,

$$J = \begin{vmatrix} \dfrac{\partial x_1}{\partial \alpha_1}(2\pi, 0, 0, 0) - 1 & \dfrac{\partial x_1}{\partial \alpha_2}(2\pi, 0, 0, 0) \\ \dfrac{\partial x_2}{\partial \alpha_1}(2\pi, 0, 0, 0) & \dfrac{\partial x_2}{\partial \alpha_2}(2\pi, 0, 0, 0 - 1 \end{vmatrix} \neq 0. \tag{9}$$

This determinant is known as the Jacobian of the system (7), and it is well known from the implicit function theorem that its nonvanishing is a sufficient condition for (7) having a solution. For future application it is convenient to introduce a short hand symbolism, and we let

$$X_\alpha = \left(\frac{\partial x_i}{\partial \alpha_j}\right), \tag{10}$$

so that X_α is a matrix whose (i, j) element is $\partial x_i/\partial \alpha_j$. Now (9) is replaced by

$$J = | X_\alpha(2\pi, 0, 0) - I | \neq 0.$$

We now wish to establish that $J \neq 0$. To do so, we shall return to (5), which is of the general form

$$X'(z, \alpha, \mu) = F(X(z, \alpha, \mu), z, \mu),$$

and differentiate the preceding equation with respect to all the components of α. This leads to using the symbolism of (10) :

$$X_\alpha'(z, \alpha, \mu) = F_X(X(z, \alpha, \mu), z, \mu)\, X_\alpha(z, \alpha, \mu).$$

We now let $\mu = \alpha = 0$, so that

$$X_\alpha'(z, 0, 0) = F_X(X(z, 0, 0), z, 0)\, X_\alpha(z, 0, 0). \tag{11}$$

The equation (11) is known as the equation of first variation. It is a linear differential equation for the matrix $X_\alpha(z, 0, 0)$, and its coefficient is periodic in time, since

$$F_X(X(z + 2\pi, 0, 0), z + 2\pi, 0) = F_X(X(z, 0, 0), z, 0)$$

In particular for (5), (11) becomes

$$X_\alpha'(z, 0, 0) = \begin{pmatrix} 0 & 1 \\ -\omega^2 & 0 \end{pmatrix} X_\alpha(z, 0, 0), \tag{12}$$

and from

$$X(0, \alpha, \mu) = X(0, 0, 0) + \alpha,$$

we obtain

$$X_\alpha(0, 0, 0) = I. \tag{13}$$

By solving (12) and (13), we find that

$$X_\alpha(z, 0, 0) = \begin{pmatrix} \cos \omega z & \dfrac{\sin \omega z}{\omega} \\ -\omega \sin \omega z & \cos \omega z \end{pmatrix},$$

so that

$$J = | X_\alpha(2\pi, 0, 0) - I | = \begin{vmatrix} \cos 2\pi\omega - 1 & \dfrac{\sin 2\pi\omega}{\omega} \\ -\omega \sin 2\pi\omega & \cos 2\pi\omega - 1 \end{vmatrix}$$

$$= 4 \sin^2 \pi\omega.$$

Clearly, if ω is not an integer, then $J \neq 0$, so that (8) is solvable. In this case (5) has a periodic solution with initial values $X(0, 0, 0) + \alpha$, where α is the solution of (8).

The fact that $J \neq 0$ could be viewed in a slightly different light. In order for (11) to have a periodic solution of period 2π, we should require that

$$| X_\alpha(2\pi, 0, 0) - \lambda I | = 0$$

has at least one solution $\lambda = 1$. Therefore $J \neq 0$ is equivalent to saying that (11) has no solution of period 2π.

Although the preceding discussion shows that (5) has a family of periodic solutions depending on μ, it is far from the best or most convenient method for obtaining this solution. A more convenient approach is to introduce for X the series:

$$X = \sum_{n=0}^{\infty} X_n \mu^n,$$

or in terms of components

$$x_1 = \sum_{n=0}^{\infty} x_1^{(n)} \mu^n ,$$

$$x_2 = \sum_{n=0}^{\infty} x_2^{(n)} \mu^n ,$$

where

$$X_0 = \begin{pmatrix} x_1^{(o)} \\ x_2^{(o)} \end{pmatrix} = \begin{pmatrix} \dfrac{\sin z}{\omega^2 - 1} \\ \dfrac{\cos z}{\omega^2 - 1} \end{pmatrix}. \tag{14}$$

In these are introduced into (5), one obtains

$$\sum_0^\infty x_1^{(n)\prime} \mu^n = \sum_0^\infty x_2^{(n)} \mu^n,$$

$$\sum_0^\infty x_2^{(n)\prime} \mu^n = -\omega^2 \sum_0^\infty x_1^{(n)} \mu^n + \sin z - \mu\omega^2 \left(\sum_0^\infty x_1^{(n)} \mu^n\right)^3. \tag{15}$$

By comparing coefficients of equal powers of μ, one obtains an infinite system of linear differential equations for the $x_j^{(n)}$. For the μ^0 terms we have

$$x_1^{(0)\prime} = x_2^{(0)},$$

$$x_2^{(o)\prime} = -\omega^2 x_1^{(o)} + \sin z,$$

and these are satisfied by (14). From the μ^1 terms we find

$$x_1^{(1)\prime} = x_2^{(1)},$$

$$x_2^{(1)\prime} = -\omega^2 x_1^{(1)} - \omega^2 (x_1^{(o)})^3. \tag{16}$$

A computation leads to the solutions

$$\begin{aligned} x_1^{(1)} &= c_1 \sin \omega z + c_2 \cos \omega z \\ &\quad - \frac{\omega^2}{(\omega^2-1)^3} \left\{ \frac{3}{4} \frac{\sin z}{\omega^2 - 1} - \frac{1}{4} \frac{\sin 3z}{\omega^2 - 9} \right\}, \\ x_2^{(1)} &= c_1 \omega \cos \omega z - c_2 \omega \sin \omega z \\ &\quad - \frac{\omega^2}{(\omega^2-1)^3} \left\{ \frac{3}{4} \frac{\cos z}{\omega^2 - 1} - \frac{3}{4} \frac{\cos 3z}{\omega^2 - 9} \right\}. \end{aligned} \tag{17}$$

(It is convenient in this computation to recall that $\sin^3 z = \frac{3}{4} \sin z - \frac{1}{4} \sin 3z$.)

The initial conditions must be so selected that the solutions (17) have period 2π; that is, they must be so selected that $c_1 = c_2 = 0$. Then one finds that

$$X = \begin{pmatrix} \dfrac{\sin z}{\omega^2 - 1} \\ \dfrac{\cos z}{\omega^2 - 1} \end{pmatrix} - \begin{pmatrix} \dfrac{3}{4} \dfrac{\sin z}{\omega^2 - 1} - \dfrac{1}{4} \dfrac{\sin 3z}{\omega^2 - 9} \\ \dfrac{3}{4} \dfrac{\cos z}{\omega^2 - 1} - \dfrac{3}{4} \dfrac{\cos 3z}{\omega^2 - 9} \end{pmatrix} \frac{\omega^2}{(\omega^2 - 1)^3} \mu + \cdots,$$

$$X(0) = \begin{pmatrix} 0 \\ \dfrac{1}{\omega^2 - 1} \end{pmatrix} + \begin{pmatrix} 0 \\ \dfrac{6\omega^2}{(\omega^2 - 1)^4 (\omega^2 - 9)} \end{pmatrix} \mu + \cdots.$$

A comparison with (7) now shows that

$$\alpha = \begin{pmatrix} 0 \\ \dfrac{6\omega^2}{(\omega^2 - 1)^4 (\omega^2 - 9)} \end{pmatrix} \mu + \cdots,$$

which, of course, could have been obtained from (8), altough with considerably more work.

We are now in a position to state and prove the following theorem.

THEOREM: Consider the differential equation

$$X' = F(X, z, \mu), \tag{18}$$

where $F(X, z, \mu)$ has continuous first derivatives in the variables X, z, μ, and has period T in the variable z. We suppose that for $\mu = 0$, (18) has a solution of period T denoted by $\bar{X}$. A sufficient condition for (18) having solutions of period T for sufficiently small $|\mu|$, such that they reduce to $\bar{X}$ for $\mu = 0$, is that the equation of variation,

$$Y' = F_X(\bar{X}, z, 0)Y,$$

has no solutions of period T.

The proof is identical to the steps carried out in the preceding example (5). We seek to adjust our initial conditions in such a way that the solution is periodic. We consider the solution $X(z, \alpha, \mu)$ such that

$$X(T, \alpha, \mu) = X(0, \alpha, \mu) = \bar{X}(0) + \alpha,$$

as was done in (7). This leads to the implicit equation for $\alpha(\mu)$:

$$X(T, \alpha, \mu) - \bar{X}(0) - \alpha = 0, \tag{19}$$

and for $\mu = 0$, we have $\alpha = 0$, since

$$X(T, 0, 0) = X(0, 0, 0) = \bar{X}(0).$$

Now (19) can be approximated for small α and μ by

$$(X_\alpha(T, 0, 0) - I)\alpha = -X_\mu(T, 0, 0)\mu + \cdots . \tag{20}$$

A solution of this equation will be assured, provided the matrix coefficient of α is not singular; that is,

$$J = | X_\alpha(T, 0, 0) - I | \neq 0. \tag{21}$$

This is comparable to (9) in the example.

To verify (21), we consider

$$X'(z, \alpha, \mu) = F(X(z, \alpha, \mu), z, \mu)$$

and differentiate this system with respect to the components of α and then set $\mu = 0$.

$$\begin{aligned} X'_\alpha(z, 0, 0) &= F_X(X(z, 0, 0), z, 0)\, X_\alpha(z, 0, 0) \\ &= F_X((\bar{X}0), z, 0)\, X_\alpha(z, 0, 0). \end{aligned} \tag{22}$$

This equation, the equation of variation, is a linear differential equation with periodic coefficients. Its solutions will be of the form (3), and the terms e^{cz} depend on the roots of the equation

$$| X_\alpha(T, 0, 0) - \lambda I | = 0. \tag{23}$$

By hypothesis, (22) has no solutions of period T, so that $\lambda = 1$ cannot satisfy (23). Therefore $J \neq 0$. Thus it has been shown that by suitable initial conditions, depending on μ, (18) will have solutions of period T.

An important corollary to the theorem can be deduced from (23).

COROLLARY: The periodic solutions obtained in the preceding theorem are asymptotically stable if for all roots of (23), $| \lambda | < 1$.

The proof follows from the fact that when all $| \lambda | < 1$, all solutions of (22) tend to zero with increasing z.

An important subcase of (18) arises in the case of autonomous differential equations. In these cases the right side of (18) does not depend explicitly on z. Then

$$X' = F(X, \mu). \tag{24}$$

One interesting feature of this type of equation is that if X is a solution of (24), then X' satisfies the equation of variation. This follows from a differentiation of (24),

$$X'' = F_X(X, \mu)X'.$$

In particular, if $\bar{X}$ is a solution of (24) of period T, then $\bar{X}'$ is a periodic solution of the equation of variation. Clearly, the theorem does not apply here, since it was assumed in the theorem that the equation of variation had no solution of period T. But nevertheless the existence of a family of periodic solutions depending on μ can be proved with some slight modifications.

THEOREM: Consider the autonomous differential equation

$$X' = F(X, \mu), \tag{24}$$

where $F(X, \mu)$ has continuous first derivatives in the variables X, μ. We suppose that for $\mu = 0$, (24) has a solution of period T_0 denoted by $\bar{X}$. A sufficient condition for (24) having periodic solutions of period $T(\mu)$ (here T has to become a function of μ) such that $T(0) = T_0$, for sufficiently small $| \mu |$, which reduce to $\bar{X}$ for $\mu = 0$, is that the equation of variation

$$Y' = F_X(\bar{X}, \mu)Y$$

has a fundamental matrix solution $\Phi(z)$ whose eigenvalue equation

$$| \Phi(T_0) - \lambda\Phi(0) | = 0$$

has $\lambda = 1$ as a simple root.

The proof is similar to the preceding proof but we let

$$T(\mu) = T_0 + \tau(\mu).$$

Then we try to adjust our initial values so that as μ varies, we still have periodic solutions. We consider the solution $X(z, \alpha, \mu)$ such that

$$X(T_0 + \tau, \alpha, \mu) = X(0, \alpha, \mu) = \bar{X}(0) + \alpha,$$

so that we have, instead of (19),

$$X(T_0 + \tau, \alpha, \mu) - \bar{X}(0) - \alpha = 0, \tag{25}$$

and for $\mu = 0$, we have $\alpha = \tau = 0$, since

$$X(T_0, 0, 0) = \bar{X}(0).$$

Now (25) is expanded for small τ, α, μ, so that

$$X_\tau(T_0, 0, 0)\tau + X_\alpha(T_0, 0, 0)\alpha + X_\mu(T_0, 0, 0)\mu - \alpha + \cdots = 0.$$

Then (20) is replaced by

$$(X_\alpha(T_0, 0, 0) - I)\alpha = -X_\mu(T_0, 0, 0)\mu - X_\tau(T_0, 0, 0)\tau + \cdots. \tag{26}$$

In this case

$$J = | X_\alpha(T_0, 0, 0) - I | = 0,$$

since by hypothesis the equation,

$$| X_\alpha(T_0, 0, 0) - \lambda I | = 0 \tag{27}$$

has a simple root at $\lambda = 1$. We now select τ so as to make (26) a compatible system. Once τ has been suitably selected, (26) can be solved for α and the existence of periodic solutions will be guaranteed. As before, we give an immediate corollary.

COROLLARY: The periodic solutions obtained in the preceding theorem are asymptotically stable if for $n - 1$ roots of (27) we have $| \lambda | < 1$. (For the *nth* root we have, of course, $\lambda = 1$.)

Systems of the type (18) or (24), for which the equations of variation do not satisfy the hypotheses required by the theorems, are known as degenerate periodic systems. Unfortunately these degenerate systems come up most frequently in practical applications. In the next chapter some second-order degenerate systems will be examined.

Problems

1. Discuss the uniqueness of the solutions of

(a) $y' = \sqrt{1 - y^2}$, $\quad y(0) = a, \quad -1 < a \leqslant 1, \quad z \geqslant 0.$

(b) $y' = \sqrt{1 - y^2}$, $\quad y(0) = -1, \quad z \geqslant 0.$

(c) $y'^2 + y^2 = 1, \quad y(0) = a, \quad -1 \leqslant a \leqslant 1, \quad z \geqslant 0.$

2. How many solutions does

$$y(y')^2 = z, \qquad y(0) = 0$$

have?

3. Show that $F(X, z)$ satisfies a Lipschitz condition if it has bounded first partial derivatives with respect to the x_i, the components of X.

4. If the series for $f(x, z)$ in (1) of Section 6.2 converges for $|x| \leqslant r$, $|z| \leqslant \rho$, show that (5) can be replaced by

$$x' = \frac{M}{\left(1 - \dfrac{x}{r}\right)\left(1 - \dfrac{z}{\rho}\right)}.$$

Trace the consequences of this in (8).

5. Show that the solution of

$$x' = 1 + z^4 + x^2, \qquad x(0) = 0,$$

cannot be bounded for all z.

6. Show that the dominating function in (21) of Section 6.2 is indeed a dominating function.

7. Find a series solution for

$$y'' = y^2, \qquad y(0) = 0.$$

8. Find an analytic solution for

$$4zy'' + 2y' = y^2, \qquad y(0) = 1.$$

Does one initial condition determine a unique solution? Explain.

9. The nonlinear equation

$$y' = a(z) + b(z)y + c(z)y^2$$

is known as the Riccati equation. Show that the substitution

$$y = -\frac{1}{c(z)}\frac{u'}{u}$$

reduces it to a second-order linear differential equation in u.

10. Consider the equation

$$X' = AX + e^{pz}C,$$

where A and C are constant matrices. Show that, if $p > \operatorname{Re} \lambda_i$ for all eigenvalues of A, then

$$\lim_{z\to\infty} e^{-pz}X = -(A - pI)^{-1}C.$$

11. Discuss the stability of the solution $x = y = 0$ of the system,

$$x' = -2x - 3y + x^5,$$
$$y' = x + y - y^2.$$

12. Discuss the stability of the solution $x = y = w = 0$ of the system,

$$x' = x - y - w,$$
$$y' = x + y - 3w,$$
$$w' = x - 5y - 3w.$$

13. For which values of α is the solution $x = y = w = 0$ of the following system stable?

$$x' = \alpha x - y,$$
$$y' = \beta y - w,$$
$$w' = \alpha w - x.$$

14. Discuss the stability of the solution $x = y = 0$ of the system

$$x' = x + e^y - \cos y,$$
$$y' = 3x - y - \sin y.$$

15. Study the periodic solutions of the equation

$$x' = ax + \sin z + \mu x^2$$

via the perturbation technique. Discuss the stability of the solution.

16. Consider the system

$$X' = AX + F(X, z) + \mu G(z),$$

where

$$F(X, z + 2\pi) = F(X, z),$$
$$G(z + 2\pi) = G(z),$$

and

$$\|F_X(0, z)\| = 0.$$

For $\mu = 0$, a periodic solution of period 2π exists, but

$$X' = AX$$

has no solution of period 2π. Show that a family of solutions for small μ exists.

17. Construct a Lyapunov function for

$$X' = \begin{pmatrix} 0 & 1 \\ -4 & -c \end{pmatrix} X, \qquad c > 0.$$

Show that a suitable function $\phi(|X|)$ exists such that

$$V \geqslant \phi(|X|).$$

18. Indicate the generalization of formula (4) of Section 6.4 to matrices with multiple eigenvalues.

19. Show that

$$u'' + \mu c u' + \omega^2(u + \mu u^3) = \sin z,$$

where $c > 0$, has periodic solutions of period 2π for small μ if ω is not an integer.

20. Construct a Lyapunov function for the stable system

$$X' = AX,$$

where A is symmetric.

7 Two-Dimensional Systems

7.1 AUTONOMOUS SYSTEMS

In many physical problems x denotes the position or state of some system, x' the velocity, and x'' the acceleration. Newton's law of motion relates the acceleration of some particle to the forces acting on it. This leads to differential equations of the form

$$x'' = f(x, x', z),$$

where z denotes the time. In this section we shall restrict ourselves to autonomous systems; that is, equations of the form

$$x'' = f(x, x'), \tag{1}$$

where z does not appear explicitly. By letting $x' = v$, we can rewrite (1) as a system of first-order equations:

$$\begin{aligned} x' &= v, \\ v' &= f(x, v). \end{aligned} \tag{2}$$

It is easy to reduce this system to a single first-order differential equation. Since both x and v are functions of z, we can try to eliminate z and think of v as a function of x. Then

$$\frac{dv}{dx} = \frac{\dfrac{dv}{dz}}{\dfrac{dx}{dz}} = \frac{f(x, v)}{v}. \tag{3}$$

Clearly, (3) is a first-order differential equation where x is the independent variable and v is the dependent variable. Once v has been determined from (3) as a function of x, one can return to the first equation in (2), which is

$$\frac{dx}{dz} = v(x),$$

and solve this equation to determine x as a function of z.

EXAMPLE. Solve the differential equation

$$x'' + x = 0.$$

We can reduce the equation to the form (2):

$$x' = v,$$
$$v' = -x,$$

and obtain, as in (3),

$$\frac{dv}{dx} = \frac{-x}{v}.$$

The preceding equation can be rewritten in the form

$$v\,dv + x\,dx = 0,$$

from which one obtains, by integration,

$$v^2 + x^2 = c^2,$$

where c^2 is a constant of integration. Then

$$x' = v = \sqrt{c^2 - x^2}.$$

This equation can be rewritten as

$$\frac{dx}{\sqrt{c^2 - x^2}} = dz,$$

and integration leads to

$$\sin^{-1}\frac{x}{c} = z + z_0,$$

where z_0 is another constant of integration. We finally obtain for x:

$$x = c \sin (z + z_0).$$

If the physical system represented by (2) has one or more equilibrium positions, they can be characterized as those solutions x, which are constants. Then both v and v' vanish and the equilibrium positions are solutions of the equation

$$f(x, 0) = 0.$$

An inspection of (3) shows that near the equilibrium positions, both the numerator and the denominator of the right side become small so that the expression becomes indeterminate. At such a point, the existence theorems of the preceding chapter fail, and a separate specific investigation of the equation must be made.

We shall now turn to a somewhat more general problem. The system to be discussed will be

$$\begin{aligned} y' &= P(x, y), \\ x' &= Q(x, y), \end{aligned} \tag{4}$$

where the prime denotes differentiation with respect to z. As in (3) we can think of y as a function of x and obtain

$$\frac{dy}{dx} = \frac{P(x, y)}{Q(x, y)}. \tag{5}$$

We shall assume that there is a point (x_0, y_0) at which both P and Q vanish simultaneously. Without loss of generality we can assume that (x_0, y_0) is the origin of our coordinate system, since we can otherwise translate the coordinates to produce this situation. Furthermore we shall assume that near the origin we can represent P and Q in the form

$$\begin{aligned} P(x, y) &= ay + bx + P_1(x, y), \\ Q(x, y) &= cy + dx + Q_1(x, y), \end{aligned}$$

where the remainder terms P_1 and Q_1 are small in the sense that

$$\lim_{\substack{x \to 0 \\ y \to 0}} \frac{P_1(x, y)}{|x| + |y|} = 0,$$

and similarly for Q_1.

It seems plausible that the behavior of the solutions of (4) and (5) should be similar to that of

$$\begin{aligned} y' &= ay + bx, \\ x' &= cy + dx, \end{aligned} \tag{6}$$

and

$$\frac{dy}{dx} = \frac{ay + bx}{cy + dx}, \tag{7}$$

near the origin. First we shall discuss (6) and (7) and then take up the question of how the solutions of (4) and (5) compare to those of (6) and (7). It is somewhat more advantageous to work with (6) rather than (7), since the former is a linear system, whereas (7) is not linear. We rewrite (6) in the form

$$X' = \begin{pmatrix} a & b \\ c & d \end{pmatrix} X, \tag{8}$$

and observe that this system can always be solved explicitly. The nature of the solutions depends strongly on the eigenvalues of the matrix coefficient.

First we shall consider the case where both eigenvalues are real. Then

$$\begin{vmatrix} a - \lambda & b \\ c & d - \lambda \end{vmatrix} = \lambda^2 - (a + d)\,\lambda + (ad - bc)$$

and

$$(a + d)^2 - 4(ad - bc) = (a - d)^2 + 4bc \geqslant 0,$$

and we shall disregard the case where $(ad - bc) = 0$. Suppose now that the matrix can be diagonalized. Then we must be able to find a nonsingular matrix T such that

$$T^{-1} \begin{pmatrix} a & b \\ c & d \end{pmatrix} T = \begin{pmatrix} \lambda_1 & 0 \\ 0 & \lambda_2 \end{pmatrix}$$

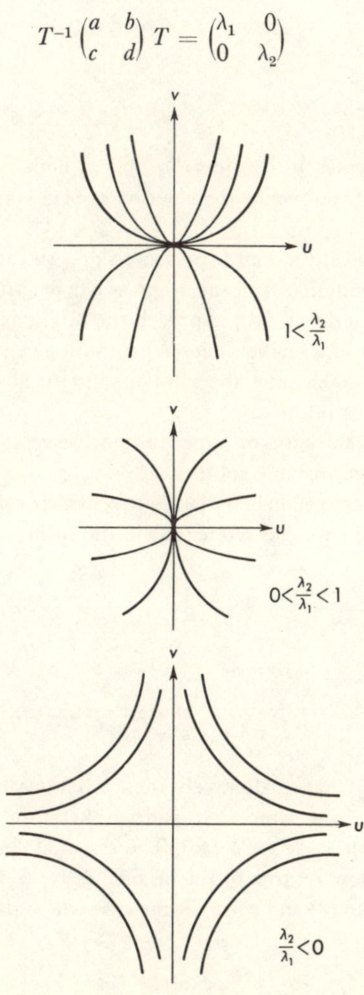

FIG. 7.1

Then the substitution

$$X = TY$$

carries (8) into

$$Y' = \begin{pmatrix} \lambda_1 & 0 \\ 0 & \lambda_2 \end{pmatrix} Y. \tag{9}$$

If we denote the components of Y as u and v, we find that

$$\begin{aligned} u &= c_1 e^{\lambda_1 z}, \\ v &= c_2 e^{\lambda_2 z}, \end{aligned} \tag{10}$$

and after elimination of z,

$$v = \kappa u^{\lambda_2/\lambda_1}. \tag{11}$$

There are three subcases to consider: λ_1 and λ_2 both negative, both positive, or of opposite sign. If we display the solution of (11) graphically, we obtain the types of curves given in Fig. 7.1.

When both eigenvalues are of the same sign, we refer to the solution as a node and refer to the singularity at the origin as a nodal point. If both eigenvalues are negative, both u and v in (10) approach the origin as z approaches infinity. In this case we speak of a stable node. When both eigenvalues are positive, we speak of an unstable node, since the solutions in (10) now move away from the origin as z approaches infinity.

When the eigenvalues are of opposite sign, we refer to the singularity as a saddle. This also is an unstable solution.

In case we cannot diagonalize the matrix, which can happen if the eigenvalues are not distinct, we can reduce (8) to the form

$$Y' = \begin{pmatrix} \lambda & 0 \\ 1 & \lambda \end{pmatrix} Y.$$

Here we find that

$$\begin{aligned} u &= c_1 e^{\lambda z}, \\ v &= (c_1 z + c_2) e^{\lambda z}. \end{aligned}$$

One can easily verify that qualitatively these solutions behave as those of (10) when λ_1 and λ_2 are of the same sign, and we therefore have a stable node for $\lambda < 0$ and an unstable node for $\lambda > 0$. The essential character of the solutions is preserved if one now returns to the original (x, y) coordinate system.

We now return to (8) and consider the case where the eigenvalues are complex conjugates, say,

$$\begin{aligned} \lambda_1 &= \mu_1 + i\mu_2, \\ \lambda_2 &= \mu_1 - i\mu_2. \end{aligned}$$

Then, as was shown in Section 6.4, it must be possible to find a real matrix T such that

$$T^{-1}\begin{pmatrix} a & b \\ c & d \end{pmatrix} T = \begin{pmatrix} \mu_1 & \mu_2 \\ -\mu_2 & \mu_1 \end{pmatrix},$$

and the substitution

$$X = TY$$

transforms (8) into

$$Y' = \begin{pmatrix} \mu_1 & \mu_2 \\ -\mu_2 & \mu_1 \end{pmatrix} Y. \tag{12}$$

To analyze this system, it is most convenient to introduce polar coordinates

$$u = \rho \cos \theta,$$
$$v = \rho \sin \theta,$$

so that, instead of (12), we obtain

$$\begin{aligned} \rho' &= \mu_1 \rho, \\ \theta' &= -\mu_2. \end{aligned} \tag{13}$$

This yields the solutions

$$\rho = c_1 e^{\mu_1 z},$$
$$\theta = -\mu_2 z + c_2.$$

These are clearly spirals. Whether they spiral in a clockwise or counterclockwise direction depends on the sign on μ_2. For $\mu_1 < 0$ we have a stable spiral, since ρ approaches the origin as z approaches infinity. For $\mu_1 > 0$ we have an unstable spiral. Figure 7.2 illustrates the case $\mu_1 < 0$, $\mu_2 < 0$.

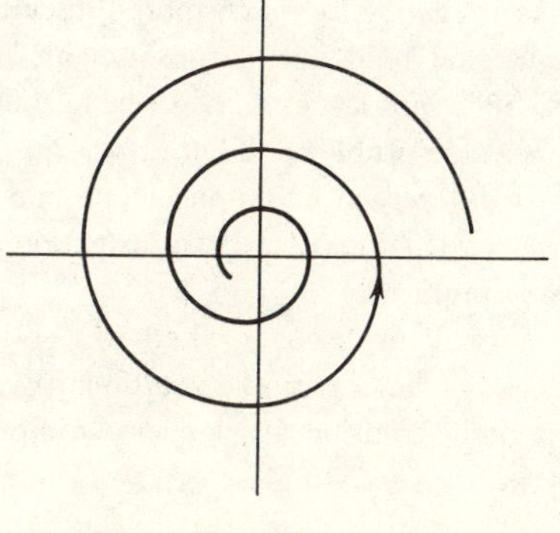

FIG. 7.2

When $\mu_1 = 0$, we have a circle, and we refer to this type of solution as a center. Here again the essential character of the solution is preserved if we return to the x, y coordinates.

We shall now rewrite (4) in the form

$$y' = ay + bx + P_1(x, y),$$
$$x' = cy + dx + Q_1(x, y),$$

or as a system:

$$X' = \begin{pmatrix} a & b \\ c & d \end{pmatrix} X + F(X); \qquad \left(F(X) = \begin{pmatrix} P_1 \\ Q_1 \end{pmatrix}\right). \tag{14}$$

By the hypotheses imposed on P_1 and Q_1, we can assert that

$$\lim_{\|X\| \to 0} \frac{\| F(X) \|}{\| X \|} = 0.$$

It follows from the stability and instability theorems of Chapter 6 that whenever (8) is stable (unstable) (14) is also stable (unstable).

In case (8) is a center, that is, $\mu_1 = 0$ in (13), no simple conclusion can be drawn with regard to the stability of (14). As a matter of fact the stability or instability of (14) depends completely on the nature of the nonlinear term.

7.2 LIMIT CYCLES

The preceding section tells us nothing about the solution of an autonomous equation except in the neighborhood of certain types of singular points. But in many problems one needs information about global solutions as distinguished from local solutions. Local properties, that is, properties holding in the neighborhood of points, can usually be handled by a variety of approximation techniques. These have been amply illustrated in the course of this book by series methods and in stability analyses.

Global properties are generally much more difficult to establish. These are properties of solutions that hold over large domains of the independent and dependent variables. Such properties have been encountered in Chapter 4 in the study of boundary value problems. There it was important to relate properties of a solution of a differential equation at one end point of an interval to properties at the other end point. Clearly, information about the solution over the whole interval was required.

In general it is difficult to discover whether a given system of differential equations does or does not have periodic solutions. Even if one can prove the existence of such a periodic solution, it is generally impossible to find that solution explicitly. Only in the very simplest cases can one hope to obtain explicit results.

EXAMPLE. Solve

$$x' = \frac{x}{(x^2 + y^2)^{1/2}}(1 - x^2 - y^2) - y,$$

$$y' = \frac{y}{(x^2 + y^2)^{1/2}}(1 - x^2 - y^2) + x.$$

The foregoing equation appears at first glance to be hopeless. However, it was carefully selected, and rewriting it in polar coordinates simplifies it greatly. Then

$$x = r \cos \theta,$$

$$y = r \sin \theta,$$

and one obtains

$$r' = 1 - r^2,$$

$$\theta' = 1.$$

Clearly, $r = 1$ is a solution, and one can verify that it is an orbitally stable solution. Therefore the original equation has the periodic solution

$$x = \cos (t + c),$$

$$y = \sin (t + c).$$

Here c is an arbitrary constant of integration. Clearly, this periodic solution is unique except for the choice of c, which involves merely a translation in time.

A solution of this type is known as a limit cycle. What made it easy to solve was that the equation had a simple character in polar coordinates. However, this is no reason to suppose that this artifice will work in general.

A limit cycle is a closed orbit in the (x, y) plane, such that no other closed orbit can be found arbitrarily close to it. We shall now turn our attention to a special class of equations for which the existence of an orbitally stable, unique limit cycle can be established.

THEOREM: Consider the differential equation

$$x'' + f(x)x' + g(x) = 0, \tag{1}$$

where $f(x)$ is a real even function and $g(x)$ is a real odd function, such that

$$xg(x) > 0, \qquad \text{for } x \neq 0,$$

$$G(x) = \int_0^x g(t)\, dt \to \infty \qquad \text{as } x \to \infty,$$

$$F(x) = \int_0^x f(t)\, dt \to \infty \qquad \text{as } x \to \infty,$$

and there exists a number $\bar{a} > 0$ such that

$$F(x) < 0, \qquad 0 < x < \bar{a},$$
$$F(x) > 0, \qquad \bar{a} < x,$$

and $F(x)$ is monotonically increasing for $x > \bar{a}$. We suppose that $f(x)$ and $g(x)$ are continuous.

Under the above hypotheses (1) must have a stable periodic solution. This solution is unique to within a translation in time.

It is remarkable that this theorem, which is easy to apply, is not difficult to prove, and yields precise information, was proved as recently as 1942. The proof is due to N. Levinson and O.K. Smith. We shall first illustrate it by means of an application to the van der Pol equation. This equation is given by

$$x'' + \mu(x^2 - 1)x' + x = 0 \tag{2}$$

where μ is a parameter. It arises in the study of electric oscillators. Clearly, the functions

$$G(x) = \int_0^x t\,dt = \frac{x^2}{2},$$
$$F(x) = \mu \int_0^x (t^2 - 1)\,dt = \mu\left(\frac{x^3}{3} - x\right),$$

satisfy the hypotheses of the theorem and the existence of a unique limit cycle is assured. There is essentially no simpler way to establish this than by applying the preceding general theorem.

We now turn to the proof of the theorem. But first we introduce the new variable

$$w = x' + F(x),$$

and then we obtain by differentiating the variable and inserting it in the equation

$$w' = -g(x).$$

It follows that we can replace (1) by the system

$$\begin{aligned} x' &= w - F(x), \\ w' &= -g(x). \end{aligned} \tag{3}$$

By elimination of the independent variable, we obtain the single first-order equation

$$\frac{dw}{dx} = \frac{-g(x)}{w - F(x)}. \tag{4}$$

We now introduce the function

$$U = \frac{1}{2}w^2 + G(x) \tag{5}$$

and observe that, depending on whether we treat z or x as the independent variable, we have

$$\begin{aligned}\frac{dU}{dx} &= w\frac{dw}{dx} + g(x) = \frac{-g(x)F(x)}{w - F(x)},\\ \frac{dU}{dw} &= w + g(x)\frac{dx}{dw} = F(x).\end{aligned} \tag{6}$$

Next we shall study the variation of U along a solution curve in the (x, w) plane, as in Fig. 7.3. We shall consider a set of initial points a, a', a'' on the positive w-axis and that part of the solution curve for which $x \geqslant 0$. As shown Fig. 7.3, at a point where the solution intersects the curve $w = F(x)$, the slope must

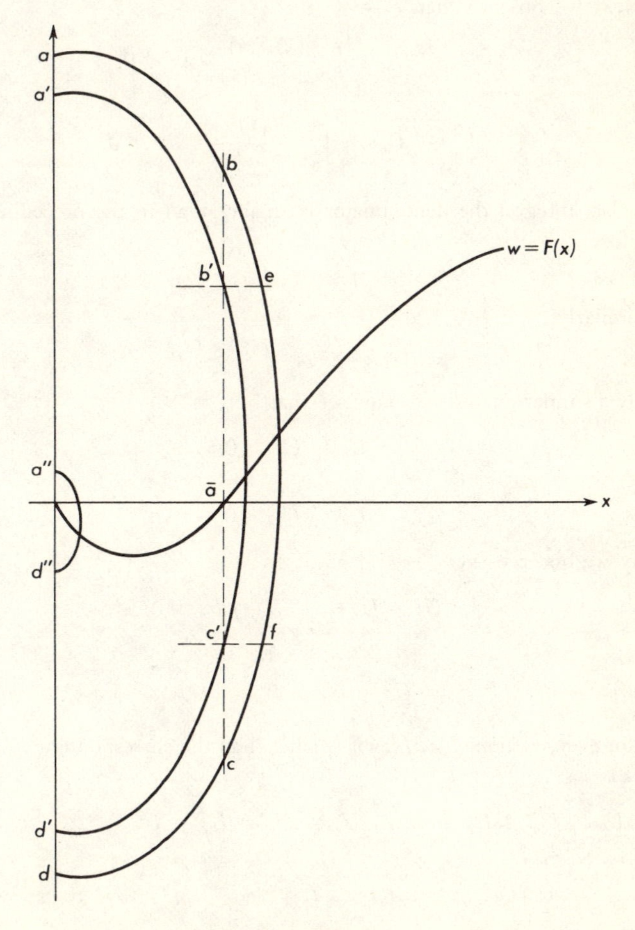

FIG. 7.3

be infinite; this is clear from (4), since at such a point the denominator vanishes. If the initial point is sufficiently small, as at a″, the maximum value attained in x lies in the range where $F(x) < 0$. But for a sufficiently large, one enters the range where $F(x) > 0$. From (6) we see that

$$U_{a''} - U_{d''} = \int_{w_{d''}}^{w_{a''}} F(x)\, dw < 0,$$

since over the interval of integration, $F(x) < 0$. By using (5) and the properties of $G(x)$, we see that

$$w_{a''}^2 - w_{d''}^2 < 0. \tag{7}$$

Next we observe that

$$U_a - U_b = \int_0^{\bar{a}} \frac{g(x)\, F(x)}{w - F(x)}\, dx < 0,$$

and

$$U_{a'} - U_{b'} = \int_0^{\bar{a}} \frac{g(x)\, F(x)}{w - F(x)}\, dx < 0.$$

In the last integral the denominator is smaller than in the preceding integral. Therefore

$$U_{a'} - U_{b'} < U_a - U_b,$$

and similarly,

$$U_{c'} - U_{d'} < U_c - U_d.$$

By a similar analysis we now see that

$$U_b - U_e > 0,$$

and

$$U_f - U_c > 0.$$

Finally we observe that

$$U_e - U_f = \int_{w_f}^{w_e} F(x)\, dw > 0,$$

and

$$U_e - U_f > U_{b'} - U_{c'}$$

since for every w along $b'c'$, $F(x)$ is smaller than the corresponding $F(x)$ along ef. Then

$$\begin{aligned} U_a - U_d &= U_a - U_b + U_b - U_e + U_e - U_f + U_f - U_c + U_c - U_d \\ &> U_a - U_b + U_e - U_f + U_c - U_d \\ &> U_{a'} - U_{b'} + U_{b'} - U_{c'} + U_{c'} - U_{d'} \\ &= U_{a'} - U_{d'}. \end{aligned} \tag{8}$$

It follows from (7) that for sufficiently small w_a, $U_a - U_d < 0$. But for sufficiently large a, $U_a - U_d$ is an increasing function of w_a, as seen in (8). We also see that as w_a increases, $U_a - U_b$, $U_b - U_c$, $U_c - U_d$ are increasing functions. $U_a - U_b$ and $U_c - U_d$ are bounded above, since they are negative. $U_b - U_c$ can grow beyond all bound, since

$$U_b - U_c = \int_{w_c}^{w_b} F(x)\, dw,$$

and for large w_a, $F(x)$ will be large along bc. Since $U_a - U_d$ is a continuous function, negative initially and subsequently monotonically increasing beyond all bound, there must be precisely one w_a for which

$$U_a - U_d = 0,$$

or equivalently, $w_a^2 = w_d^2$.

The functions $g(x)$ and $F(x)$ in (4) are odd functions. If follows that if w is replaced by $-w$ and x by $-x$, (4) remains unchanged. Therefore the solution for $x < 0$, which starts at w_d, must return to w_a. The resultant solution therefore returns to its initial point and must be periodic. In the (x, w) plane, the limit cycle is unique. In the time domain any two solutions must correspond to the same limit cycle in the (x, w) plane and can therefore differ only in a displacement of the time origin.

We still have to establish the orbital stability of the limit cycle. To do so, we observe that if w_a is too large, $U_a - U_d > 0$, so that $w_d^2 < w_a^2$, and the solution must move inward toward the limit cycle. If, on the other hand, w_a is too small, $w_d^2 > w_a^2$ and the solution moves outward. In either case the solution will approach the limit cycle asymptotically.

At the beginning of this section we mentioned that it is in general very difficult to establish the existence of a limit cycle. One theorem of a rather general character that yields a sufficient condition for a limit cycle is the Poincaré-Bendixson theorem.

THEOREM: Consider the differential equation

$$\frac{dy}{dx} = \frac{P(x, y)}{Q(x, y)}, \tag{9}$$

where P and Q are continuous functions for all x and y. Suppose $x(z)$, $y(z)$ represent a solution curve in the (x, y) plane, which stays in a bounded region as $z \to \infty$, and which contains no singular points of (9). Then there exists at least one limit cycle.

In spite of apparent simplicity of the theorem it is generally very difficult to apply, and we shall refrain from giving a proof.

A theorem that is sometimes useful for the purpose of establishing the nonexistence of a limit cycle is due to Bendixson.

THEOREM: If P and Q in (9) have continuous partial derivatives in a domain bounded by a simple curve C, and if $\partial P/\partial y + \partial Q/\partial x$ is of constant sign in that domain, then (9) can have no limit cycle in that domain.

The proof is based on an application of Gauss' theorem, which states that

$$\iint_A \left(\frac{\partial P}{\partial y} + \frac{\partial Q}{\partial x}\right) dx\, dy = \int_S (Q\, dy - P\, dx), \tag{10}$$

where on the left we have an integral over the area A and on the right a line integral along the boundary S of A.

If we assume that (9) has a limit cycle S, then clearly on S,

$$Q\, dy - P\, dx = 0,$$

so that the right side of (10) must vanish. But by hypothesis the left side of (10) cannot vanish. Clearly, no limit cycle can exist. Incidentally, (10) furnishes a necessary condition for S's being a limit cycle, since then the left side must vanish.

EXAMPLE. The differential equation

$$\frac{dy}{dx} = \frac{ay + f(x)}{bx + g(y)},$$

where $f(x)$ and $g(y)$ are regular for all x and y, can have no limit cycles if $a + b \neq 0$. The proof, of course, follows from the fact that

$$\frac{\partial P}{\partial y} + \frac{\partial Q}{\partial x} = a + b.$$

In the preceding chapter some general conclusions were drawn regarding autonomous systems. Since a second-order system can be expected to be a little simpler than a general n'th-order system, one should expect that some of the conclusions can be strengthened. We suppose that the system

$$X' = F(X), \tag{11}$$

where $F(X)$ has continuous derivatives, has a periodic solution $\bar{X}$, of period T, say, and we shall now observe what happens if we perturb it slightly; that is, we let

$$X = \bar{X} + \epsilon Y,$$

where ϵ is small. Then (11) becomes

$$\epsilon Y' = F(\bar{X} + \epsilon Y) - \bar{X}'.$$

If we now expand the terms on the right for small ϵ, we obtain

$$F(\bar{X} + \epsilon Y) = F(\bar{X}) + \epsilon F_X(\bar{X})Y + \epsilon R,$$

where R is a suitable remainder term. Then, since $\bar{X}$ satisfies (11), we obtain

$$Y' = F_X(\bar{X})Y, \tag{12}$$

if we neglect the remainder term; (12) is a linear differential equation with periodic coefficients, and in Chapter 6 was called the equation of variation. From our general conclusions of Chapter 5 we know that (12) must have solutions of the form

$$Y_i = P_i(z)e^{\lambda_i z},$$

where $P_i(z)$ is periodic. The λ_i were determined by constructing a fundamental matrix solution Φ of (12) and then

$$e^{\lambda_i T} = \mu_i,$$

where the μ_i were the roots of the equation

$$| \Phi(T) - \mu\Phi(0) | = 0. \tag{13}$$

We shall assume merely for the sake of simplicity that all roots of (13) are simple in the case of (12). By differentiating (11), we obtain

$$\bar{X}'' = F_X(\bar{X})\bar{X}',$$

showing that $\bar{X}'$ satisfies (12). Therefore (12) has at least one periodic solution and (13) at least one root $\mu = 1$. If all other roots of (13) are less than unity in magnitude, all solutions of (12) except X' will decay with time, showing that X is a stable solution of (11).

We shall now apply these results to the special autonomous system

$$\begin{aligned} y' &= P(x, y), \\ x' &= Q(x, y). \end{aligned} \tag{14}$$

We shall denote the periodic solution of (14) with y and x, and the equation of first variation will become

$$\begin{aligned} \eta' &= P_x(x, y)\, \xi + P_y(x, y)\, \eta, \\ \xi' &= Q_x(x, y)\, \xi + Q_y(x, y)\, \eta. \end{aligned} \tag{15}$$

By hypothesis this system has a periodic solution y', x', and a second solution such that

$$\eta(z + T) = \mu\eta(z),$$
$$\xi(z + T) = \mu\xi(z).$$

Then a fundamental matrix is given by

$$\Phi = \begin{pmatrix} \eta & y' \\ \xi & x' \end{pmatrix}.$$

Clearly,

$$\Phi(T) = \mu\Phi(0),$$

so that

$$\mu = \frac{|\,\Phi(T)\,|}{|\,\Phi(0)\,|}. \tag{16}$$

We now consider the Wronskian

$$W = |\,\Phi\,|$$

and find that

$$W' = (P_y + Q_x)W. \tag{17}$$

This could have been deduced from the general statement that if

$$\Phi' = A\Phi, \qquad W' = (\text{tr } A)W.$$

From (17) and (16) we now obtain

$$\mu = \frac{W(T)}{W(0)} = \exp\left(\int_0^T (P_y + Q_x)\, dz\right). \tag{18}$$

For orbital stability we now require that

$$|\,\mu\,| < 1,$$

so that

$$\int_0^T (P_y + Q_x)\, dz < 0. \tag{19}$$

The importance of this stability criterion is that, since an estimate of the magnitude μ is obtained, one can also determine the rate at which a solution curve approaches the limit cycle. This will be of use to us in Section 7.4.

7.3 THE POINCARÉ INDEX

This section will be devoted to a discussion of the Poincaré index, by means of which one can derive a convenient necessary condition for the existence of periodic solutions of second-order autonomous differential equations. The system

$$\begin{aligned} \frac{dy}{dz} &= P(x, y), \\ \frac{dx}{dz} &= Q(x, y), \end{aligned} \tag{1}$$

or equivalently the first-order differential equation

$$\frac{dy}{dx} = \frac{P(x, y)}{Q(x, y)}, \tag{2}$$

define a unique direction field at every point of the (x, y) plane where the right side of (2) is defined; that is, where

$$P^2 + Q^2 \neq 0.$$

One can rephrase this statement by saying that (2) has a unique solution through the initial point (x_0, y_0), provided

$$P^2(x_0, y_0) + Q^2(x_0, y_0) \neq 0. \tag{3}$$

If this is not the case, then the right side in (2) is indeterminate. In Section 7.1 we examined the solutions in cases where (3) was violated and P and Q behaved like linear functions near (x_0, y_0).

DEFINITION. Consider a simple, nonself-intersecting, closed curve C on which $P^2 + Q^2 \neq 0$. With every point of C we can associate an angle $\theta = \tan^{-1}(P/Q)$, which defines the direction in which the solution of (2) through that point moves. We now start with a definite point on C and observe how θ changes as we move along C in a counterclockwise direction. Upon returning to the starting point, θ must attain a value that differs from the initial value by an integral multiple of 2π, say, $2\pi J$. J will be called the index of the curve C.

We can easily derive an analytic expression for J. Since $2\pi J$ denotes the change in θ as we traverse the curve C, we have

$$J = \frac{1}{2\pi} \int_C d\theta = \frac{1}{2\pi} \int_C d \tan^{-1} \frac{P}{Q}. \tag{4}$$

The line integral above must be evaluated in a counterclockwise direction. Certain properties of the index follow readily from the definition.

PROPERTY 1. If C_1 and C_2 are two simple, nonself-intersecting, closed curves, with indices J_1 and J_2, respectively, such that C_1 can be continuously deformed into C_2 without crossing any singular points (that is, points where $P^2 + Q^2 = 0$), then

$$J_1 = J_2.$$

The proof relies on the fact that a line integral varies continuously with the path of integration as long as no singularity of the integrand is crossed. Since (4) takes on only integral values, on deformations of the curve C, it must either remain constant or undergo discontinuous changes. The latter can occur if and only if a singularity is crossed. Since no singularity can be encountered on deforming C_1 into C_2, J must remain constant.

PROPERTY 2. The index of a curve C that encloses no singular points must vanish.

In this case C can be deformed into an arbitrarily small curve without changing J. As the curve shrinks to a point, θ reduces to a constant so that $J = 0$.

As a consequence of Property 1 we can introduce the notion of the index of a singular point.

DEFINITION: The index of a singular point is defined as the index of any simple, nonself-intersecting, closed curve C enclosing only this one singularity and no other.

The foregoing statement yields an unambiguous definition of the index of a singularity, since if C_1 and C_2 are two curves enclosing only this one singularity, then C_1 can be deformed into C_2 without crossing another singularity, so that $J_1 = J_2$.

PROPERTY 3. If the curve C encloses a finite number of singular points, then the index of C is equal to the sum of the indices of the enclosed singularities.

We shall indicate the proof for the case where C encloses only two singularities at S_1 and S_2, as shown in Fig. 7.4.

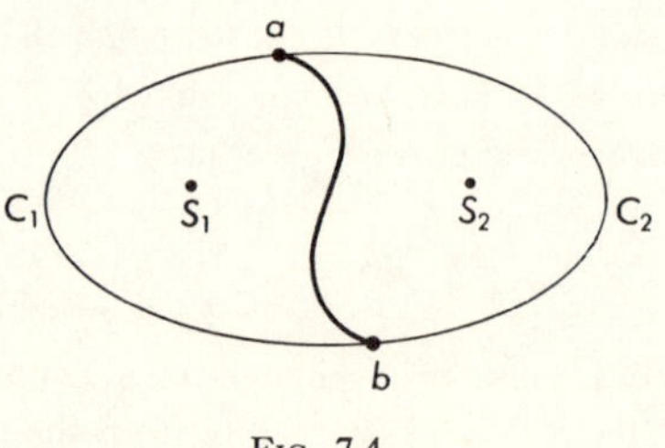

FIG. 7.4

If a and b represent two arbitrary points on C, we can connect them with an arc lying entirely inside C and separating the interior into two subdomains, one containing only S_1 and the other only S_2. We denote the new closed curves enclosing S_1 and S_2 by C_1 and C_2, respectively. If J_1 and J_2 denote the indices of S_1 and S_2, then the index of C is given by

$$\begin{aligned} J &= \frac{1}{2\pi} \int_C d \tan^{-1} \frac{P}{Q} \\ &= \frac{1}{2\pi} \int_{C_1} d \tan^{-1} \frac{P}{Q} + \frac{1}{2\pi} \int_{C_2} d \tan^{-1} \frac{P}{Q} \\ &= J_1 + J_2. \end{aligned}$$

That the sum of the integrals over C_1 and C_2 reduces to the integral over C follows from the fact that both former integrals contain, in part, integrals over the arc ab, but in opposite directions. Upon adding the two integrals, the contributions due to the arc ab cancel out, and the sum reduces to the integral over C.

PROPERTY 4. For a closed curve C that represents a periodic solution of (1), we find that $J = 1$.

In this case the direction field defined by (1) or (2) must be tangent to C at every point of C, since the curve C must satisfy (2). As we traverse C in a counterclockwise direction, the tangent vector makes precisely one counterclockwise rotation. Therefore $J = 1$.

In Section 7.1 several types of singularities were discussed. These were nodes, saddles, and spirals. We can show that the index of a saddle is -1 and that the index of a node or spiral is $+1$. To do so, we use equation (9) of Section 7.1; that is,

$$\frac{dy}{dz} = \lambda_1 y,$$

$$\frac{dx}{dz} = \lambda_2 x,$$

so that

$$J = \frac{1}{2\pi}\int_C d\tan^{-1}\frac{\lambda_1 y}{\lambda_2 x} = \frac{1}{2\pi}\int_C \frac{\lambda_1\lambda_2(x\,dy - y\,dx)}{(\lambda_1 y)^2 + (\lambda_2 x)^2}.$$

We shall consider the case where λ_1, λ_2 are real, and for C we select the ellipse

$$|\lambda_2|\,x = \cos z,$$

$$|\lambda_1|\,y = \sin z.$$

In the preceding expressions we wrote $|\lambda_1|$ and $|\lambda_2|$ to be sure that as z runs from 0 to 2π, the ellipse will be traversed in a counterclockwise direction. Then

$$J = \frac{\lambda_1\lambda_2}{|\lambda_1\lambda_2|}\frac{1}{2\pi}\int_0^{2\pi} dz = \frac{\lambda_1\lambda_2}{|\lambda_1\lambda_2|} = \begin{cases} +1, & \lambda_1\lambda_2 > 0, \\ -1, & \lambda_1\lambda_2 < 0. \end{cases}$$

For λ_1 and λ_2 of equal sign, we have a node of index $+1$; and for λ_1 and λ_2 of opposite sign, a saddle of index -1. All other cases (that is, spirals and nodes, where $\lambda_1 = \lambda_2$) can be discussed in a similar manner.

EXAMPLE 1. The van der Pol equation

$$x'' + \mu(x^2 - 1)x' + x = 0$$

can be rewritten as a system:

$$x' = v,$$

$$v' = -\mu(x^2 - 1)v - x.$$

The point $x = v = 0$ is a singular point, and a simple computation that compares the foregoing expressions to the linearized system

$$x' = v$$
$$v' = \mu v - x$$

shows it to be a spiral or node of index 1, that is unstable. Therefore a periodic solution is possible. That a limit cycle really does exist follows from the Levinson-Smith theorem.

EXAMPLE 2. The second-order equation

$$y'' + ay + by^3 = 0$$

can be rewritten as a system

$$\begin{aligned} y' &= x, \\ x' &= -ay - by^3. \end{aligned} \tag{5}$$

The solution in the (x, y) plane is easily found in this case and becomes

$$x^2 + ay^2 + \tfrac{1}{2}by^4 = K, \tag{6}$$

where K is a constant of integration. But the nature of the solution (6) and its dependence on the parameters a and b is much easier to determine by working with both (5) and (6) rather than with (6) alone. The simplest case arises when both a and b are positive. Then the only singular point of (5) is $x = y = 0$, and it is easily shown to be a center. Therefore periodic solutions are possible From (6) we now see that, qualitatively, we have solutions in the (x, y) plane as shown in Fig. 7.5.

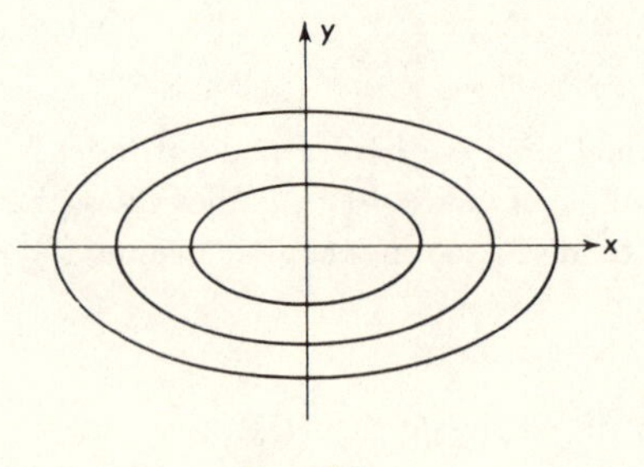

FIG. 7.5

Next we examine the case where a and b are negative. Again the point $x = y = 0$ is the only singular point. But this time we have a saddle of index

-1. Therefore no closed orbits are possible in this case. The solution through the origin is given by

$$x^2 + ay^2 + \tfrac{1}{2}by^4 = 0.$$

By solving for x, we obtain two branches

$$x = \pm \sqrt{(-a)\,y^2 + \tfrac{1}{2}(-b)\,y^4}. \tag{7}$$

These split the (x, y) plane into four regions (Fig. 7.6), and any solution that

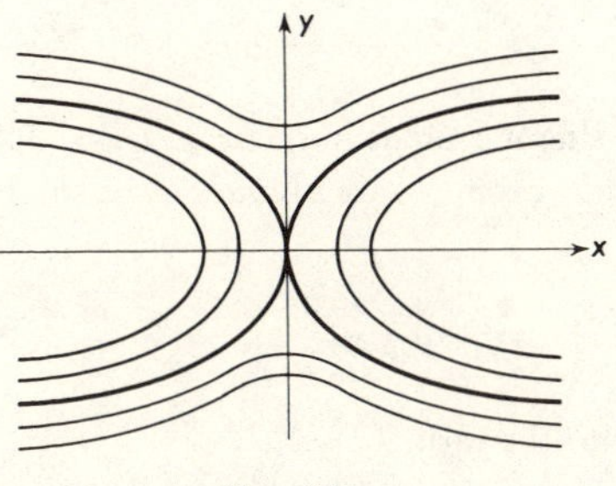

FIG. 7.6

starts in one of these must stay in it; otherwise, two solutions would intersect. But two solutions through the same point must be identical unless that point is a singular point. As we have already seen, the only solutions through the singular point are given by (7).

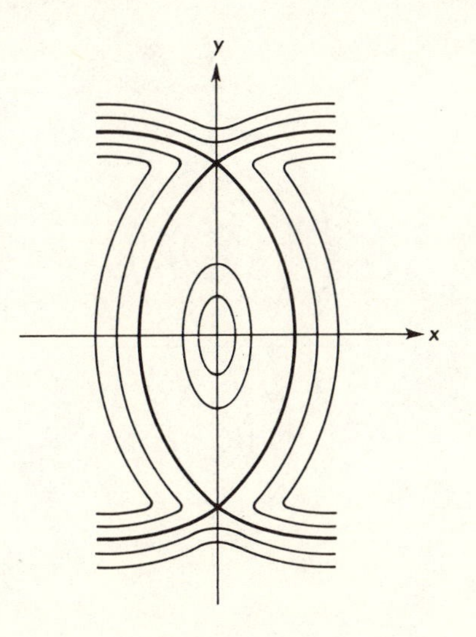

FIG. 7.7

The third case we shall discuss is the case where $b < 0 < a$ (Fig. 7.7). Then (5) has singular points at $x = 0$, $y = 0$, and at $x = 0$, $y = \pm\sqrt{-a/b}$. The origin is easily seen to be a center. To examine the other points, we let

$$y = \sqrt{-\frac{a}{b}} + w,$$

for example, so that (5) becomes, after higher powers of w have been neglected,

$$w' = x,$$
$$x' = 2aw.$$

We thus see that the singular point in question is a saddle, and similarly for the point at $y = -\sqrt{-a/b}$. From (6) we obtain for the solution, which goes through the saddle points,

$$x^2 + ay^2 + \frac{1}{2}by^4 = -\frac{a^2}{2b},$$

which can factored into the form

$$\left(x + \frac{a}{\sqrt{-2b}} - \sqrt{-\frac{b}{2}}y^2\right)\left(x - \frac{a}{\sqrt{-2b}} + \sqrt{\frac{-b}{2}}y^2\right) = 0.$$

Thus we see that the solution (6) through the saddle points has two parabolic branches. These branches split the x, y plane into five subregions. Four

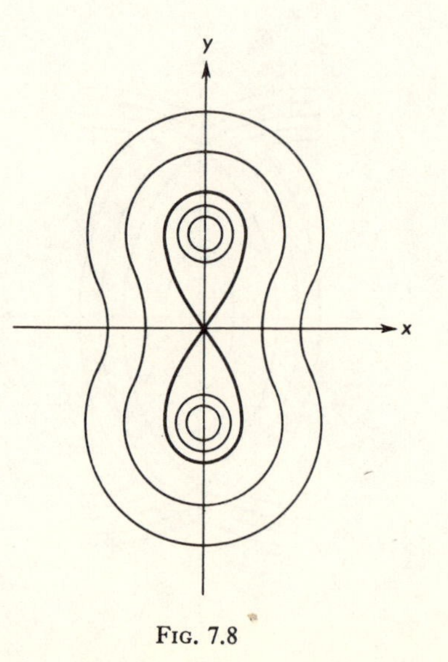

Fig. 7.8

of these are infinite in extent and contain no closed orbits. But the fifth region contains the origin, which contains a singular points of index $+1$. Here closed orbits are possible and do indeed exist. No closed orbit could contain all three singular points because its index would have to be $+1$ to qualify as a closed orbit. But the sum of the indices of the enclosed singular points is -1.

The last case we have to consider is the case $a < 0 < b$ (Fig. 7.8). This time we find that the origin $x = y = 0$ is a saddle point and the points $x = 0$, $y = \pm\sqrt{-a/b}$ are centers. The solution through the saddle is given by

$$x^2 + ay^2 + \tfrac{1}{2}by^4 = 0,$$

which has the branches

$$x = \pm y\sqrt{-a - \frac{b}{2}y^2}. \tag{8}$$

The extreme values y can attain on these branches are $y = \pm\sqrt{-2a/b}$.

It is seen that (8) splits the (x, y) plane into two finite and one infinite regions. In each finite region we find one singular point of index $+1$, and all solutions originating inside these finite regions are periodic. In this case orbits enclosing all three singular points may exist, since the sum of their indices is $+1$. Such orbits do indeed exist. As a matter of fact, for all initial points not on (8), we find closed orbits.

One final example will be offered to illustrate these methods.

EXAMPLE 3. The differential equation

$$\frac{dy}{dx} = \frac{x - x^3}{y - y^3} \tag{9}$$

can be integrated immediately, and the solutions are given by

$$y^2 - \tfrac{1}{2}y^4 - x^2 + \tfrac{1}{2}x^4 = K, \tag{10}$$

where K is a constant of integration. The equation has nine singular points at all possible combinations of the coordinates $x = 0$, $x = \pm 1$, $y = 0$, $y = \pm 1$. The singularities at $(0, 0)$, $(\pm 1, \pm 1)$ are saddles and at $(0, \pm 1)$, $(\pm 1, 0)$ are centers. The solution 10), which goes through the saddles, is given by

$$y^2 - \tfrac{1}{2}y^4 - x^2 + \tfrac{1}{2}x^4 = 0,$$

or in factored form,

$$\tfrac{1}{2}(y^2 - x^2)(1 - x^2 - y^2) = 0.$$

The solutions of these equations divide the (x, y) plane into four bounded and

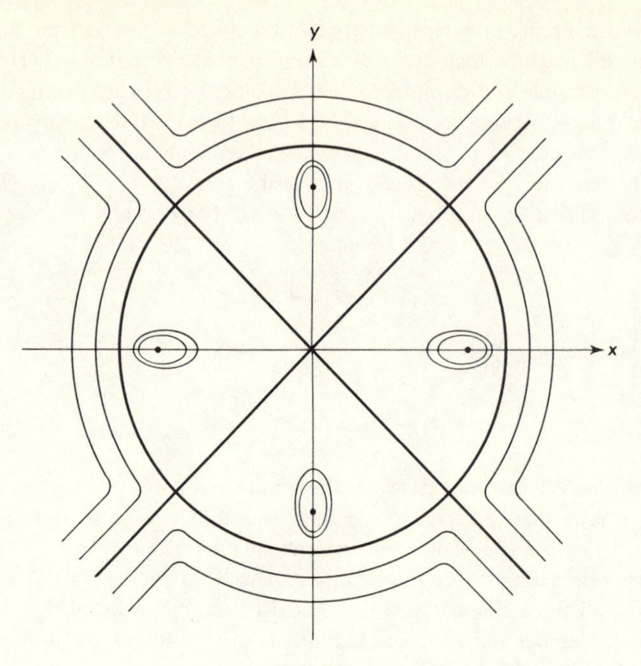

Fig. 7.9

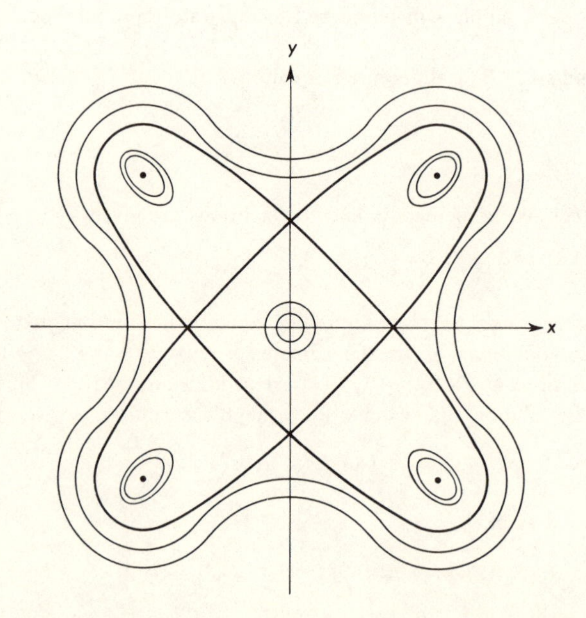

Fig. 7.10

four unbounded regions (Fig. 7.9). The centers are within the four bounded regions, and within these we have closed orbits.

As a companion equation to (9) we consider

$$\frac{dy}{dx} = -\frac{x - x^3}{y - y^3}, \tag{11}$$

whose solution is

$$y^2 - \tfrac{1}{2}y^4 + x^2 - \tfrac{1}{2}x^4 = K. \tag{12}$$

The singular points for (11) are the same as those of (9) except that at (0, 0), (± 1, ± 1), we have centers and at (0, ± 1), (± 1, 0) saddles. The solution (Fig. 7.10), which goes through the saddles, is given by

$$y^2 - \tfrac{1}{2}y^4 + x^2 - \tfrac{1}{2}x^4 = \tfrac{1}{2},$$

or in factored form

$$\frac{1}{2}(y^2 + x^2 + \sqrt{2}\,xy - 1)(y^2 + x^2 - \sqrt{2}\,xy - 1) = 0.$$

The two ellipses given by the foregoing expression divide the (x, y) plane into five bounded and one unbounded regions. Each of the bounded regions contains a center of index $+1$, and all solutions within these regions are closed orbits.

In addition we found closed orbits that lie in the unbounded region and which enclose five singularities of index $+1$ and four of index -1.

7.4 PERTURBATION THEORY OF SECOND-ORDER SYSTEMS

In this section we shall return to some problems first touched on in the preceding chapter; that is, perturbation theory of second-order equations. In particular we shall discuss the equation

$$x'' + x = \beta f(x, x', z, \beta), \tag{1}$$

where β is small and the right side is periodic in z; that is, for some T,

$$f(x, x', z + T, \beta) = f(x, x', z, \beta).$$

For $\beta = 0$, evidently (1) has solutions of period 2π. By the method of variation of parameters, (1) can be rewritten as an integral equation:

$$x = c_1 \cos z + c_2 \sin z + \beta \int_0^z \sin (z - \tau) f(x, x', \tau, \beta)\, d\tau, \tag{2}$$

from which we also see that

$$x' = -c_1 \sin z + c_2 \cos z + \beta \int_0^z \cos (z - \tau) f(x, x', \tau, \beta)\, d\tau. \tag{3}$$

The solution (2) clearly satisfies the initial conditions

$$x(0) = c_1,$$
$$x'(0) = c_2.$$

We are primarily concerned with discovering whether (1) has solutions of period T. We have a theorem that guarantees the existence of solutions for sufficiently small β, provided the equation of first variation has no solution of period T. In the case of (1), the equation of first variation is given by

$$x'' + x = 0.$$

Therefore, for $T \neq 2\pi$, everything is clear.

We shall now consider the special case where

$$f(x, x', z, \beta) = -x^3 + F \cos \omega z, \tag{4}$$
$$\omega = \omega(\beta) = 1 + \beta\omega_1 + \beta^2\omega_2 + \cdots.$$

In this case we cannot take recourse to our standard existence theorem, since for $\beta = 0$, we do have $T = 2\pi$. The equation (1) with the right side given by (4) is known as Duffing's equation. It is convenient for our analysis to introduce a new independent variable s by letting

$$s = \omega z.$$

Then we obtain, instead of (1) and (4),

$$\omega^2 \frac{d^2x}{ds^2} + x + \beta x^3 = \beta F \cos s. \tag{5}$$

Since the equation is invariant under a replacement of s by $-s$, we need seek only even solutions because the forcing function (that is, $\beta F \cos s$) is an even function. Then we can operate with initial conditions of the form

$$x(0) = A, \qquad x'(0) = 0.$$

In the absence of the βx^3 term, the solution of (5) would be given by

$$x = \left(A - \frac{\beta F}{1 - \omega^2}\right) \cos \frac{s}{\omega} + \frac{\beta F}{1 - \omega^2} \cos s.$$

By an application of the method of variation of parameters we are then led to the equations

$$x = \left(A - \frac{\beta F}{1 - \omega^2}\right) \cos \frac{s}{\omega} + \frac{\beta F}{1 - \omega^2} \cos s - \frac{\beta}{\omega} \int_0^s \sin \frac{s - \tau}{\omega} x^3 \, d\tau, \tag{6}$$

$$x' = -\frac{1}{\omega}\left(A - \frac{\beta F}{1 - \omega^2}\right) \sin \frac{s}{\omega} - \frac{\beta F}{1 - \omega^2} \sin s - \frac{\beta}{\omega^2} \int_0^s \cos \frac{s - \tau}{\omega} x^3 \, d\tau. \tag{7}$$

In general we saw that from (6) and (7) and the periodicity of x and x', we should be able to determine the initial conditions necessary for periodicity. In this problem we know that $x'(0) = 0$, from the fact that our solution must be even in time. We now require that $x'(2\pi) = 0$, which leads to, by the use of (7),

$$-\frac{1}{\omega}\left(A - \frac{\beta F}{1-\omega^2}\right)\sin\frac{2\pi}{\omega} - \frac{\beta}{\omega^2}\int_0^{2\pi}\cos\frac{2\pi - \tau}{\omega}\,x^3\,d\tau = 0. \tag{8}$$

Equation (8) is of the form

$$G(A, \omega(\beta), \beta) = 0,$$

and $\omega(\beta)$ was presumably a prescribed function of β. However, at this point we shall change our viewpoint slightly and regard A as a prescribed initial condition; we shall seek an $\omega(\beta)$ that will satisfy (8). For small β we find that

$$\omega = 1 + \beta\omega_1 + \cdots,$$

$$\sin\frac{2\pi}{\omega} = \sin 2\pi(1 - \beta\omega_1 + \cdots) = -2\pi\beta\omega_1 + \cdots,$$

$$\frac{\beta F}{1-\omega^2} = -\frac{F}{2\omega_1} + \cdots,$$

$$x = A\cos s + \cdots,$$

so that (8) can be rewritten in the form

$$\beta\left[\left(A + \frac{F}{2\omega_1}\right)2\pi\,\omega_1 - \int_0^{2\pi} A^3\cos^4\tau\,d\tau\right] + \cdots = 0. \tag{9}$$

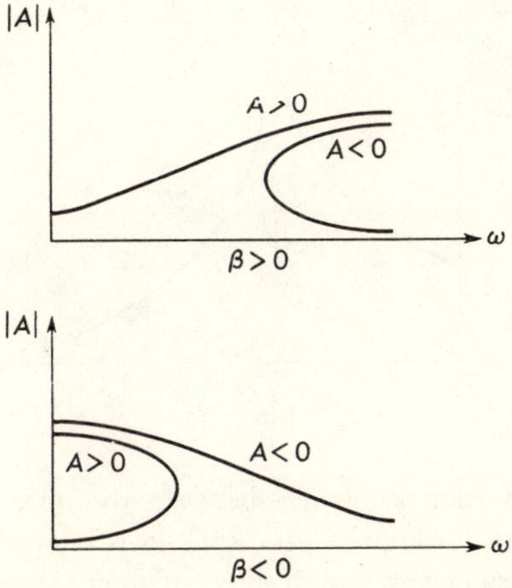

FIG. 7.11

The dots in all the foregoing expressions denote higher-order terms in β. By performing the integration in (9) and setting the bracketed term equal to zero, we obtain the following expression for ω_1:

$$\omega_1 = \frac{3}{8}A^2 - \frac{F}{2A}.$$

From this we see that ω_1 is a single-valued function of A, whereas A is a triple-valued function of ω_1. This is the basic motivation for the previous change of viewpoint. It follows that

$$\omega = 1 + \beta\omega_1 + \cdots = 1 + \beta\left(\frac{3}{8}A^2 - \frac{F}{2A}\right) + \cdots. \tag{10}$$

We can now make a graph (Fig. 7.11) of the functional relationship between A and ω. It is more convenient to choose ω as abscissa and $|A|$ as ordinate. Graphs of the type of Fig. 7.11 are often known as response curves, since they show the amplitude with which the system responds to a forcing function of frequency ω. It is interesting to compare this to the comparable curves for linear systems (Fig. 7.12):

$$x'' + x = F\cos\omega z,$$

$$x = \frac{F}{1-\omega^2}\cos\omega z,$$

so that

$$|A| = \frac{F}{|1-\omega^2|}.$$

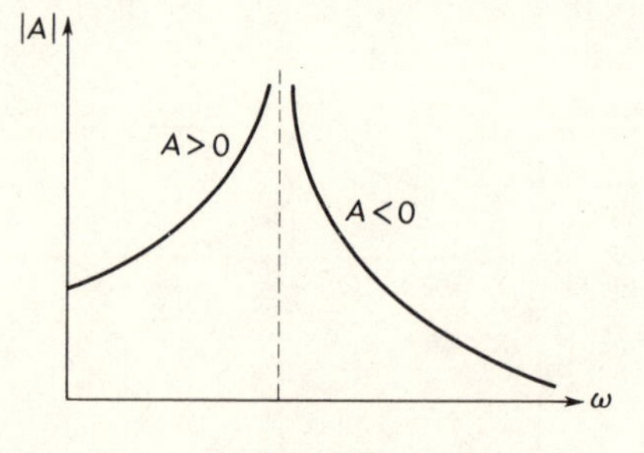

FIG. 7.12

The most significant difference between the linear and nonlinear cases is that, in the former for a given $\omega(\omega \neq 1)$, A is uniquely defined. But in the nonlinear case a given ω may lead to three different values of A. In other words, there are three possible periodic solutions.

We now turn to another problem, which will illustrate still another aspect of nonlinear differential equations. The equation to be studied now is

$$\omega^2 x'' + x + \beta x^3 = F \cos s, \tag{11}$$

which is similar to (5) except that the right side is large in the sense that it is independent of β.

In view of the fact that the right side of (11) has period 2π, one might expect to find solutions of period 2π. But it will be shown that as a consequence of the nonlinear term, it will be possible to find solutions of period 6π; that is, solutions of frequency 1/3 of that of the forcing term in (11). Such solutions are known as subharmonics. For example, the equation

$$\omega^2 x'' + x + \beta x^3 = \cos s, \tag{11$'$}$$

where

$$\omega^2 = 9\left[1 + 3\left(\frac{\beta}{4}\right)^{1/3}\right]$$

has the solution

$$x = \left(\frac{4}{\beta}\right)^{1/3} \cos \frac{s}{3}.$$

We shall therefore seek solutions of (11) of period 6π such that ω is a function of β, which for $\beta = 0$ reduces to $\omega = 3$. Since (11) is invariant under a replacement of s by $-s$, we can restrict ourselves to a search for even solutions. Then, as in (6) and (7), we have

$$x = \left(A - \frac{F}{1-\omega^2}\right) \cos \frac{s}{\omega} + \frac{F}{1-\omega^2} \cos s - \frac{\beta}{\omega} \int_0^s \sin \frac{s-\tau}{\omega}\, x^3\, d\tau, \tag{12}$$

$$x' = -\frac{1}{\omega}\left(A - \frac{F}{1-\omega^2}\right) \sin \frac{s}{\omega} - \frac{F}{1-\omega^2} \sin s - \frac{\beta}{\omega^2} \int_0^s \cos \frac{s-\tau}{\omega}\, x^3\, d\tau. \tag{13}$$

We shall now seek to determine a relationship between A and ω from the periodicity condition

$$x'(6\pi) = x'(0) = 0.$$

Letting

$$\omega = 3 + \beta\omega_1 + \cdots,$$

$$\sin \frac{6\pi}{\omega} = \frac{-2\pi\beta\omega_1}{3} + \cdots,$$

$$x = \left(A + \frac{F}{8}\right) \cos \frac{s}{3} - \frac{F}{8} \cos s + \cdots,$$

we obtain from (13),

$$\beta\left[\frac{2\pi}{9}\left(A+\frac{F}{8}\right)\omega_1-\frac{1}{9}\int_0^{6\pi}\cos\frac{s}{3}\left\{\left(A+\frac{F}{8}\right)\cos\frac{s}{3}-\frac{F}{8}\cos s\right\}^3 ds\right]+\cdots=0. \tag{14}$$

From (14) we finally obtain

$$\omega_1=\frac{9}{8}\left(A^2+\frac{AF}{8}+\frac{F^2}{32}\right),$$

so that

$$\omega=3+\beta\omega_1+\cdots=3+\frac{9\beta}{8}\left(A^2+\frac{AF}{8}+\frac{F^2}{32}\right)+\cdots. \tag{15}$$

The response curves for (15) are shown in Fig. 7.13. Some interesting conclusions can drawn from a superposition of the response curves for (10) and (15). These are shown in Fig. 7.14.

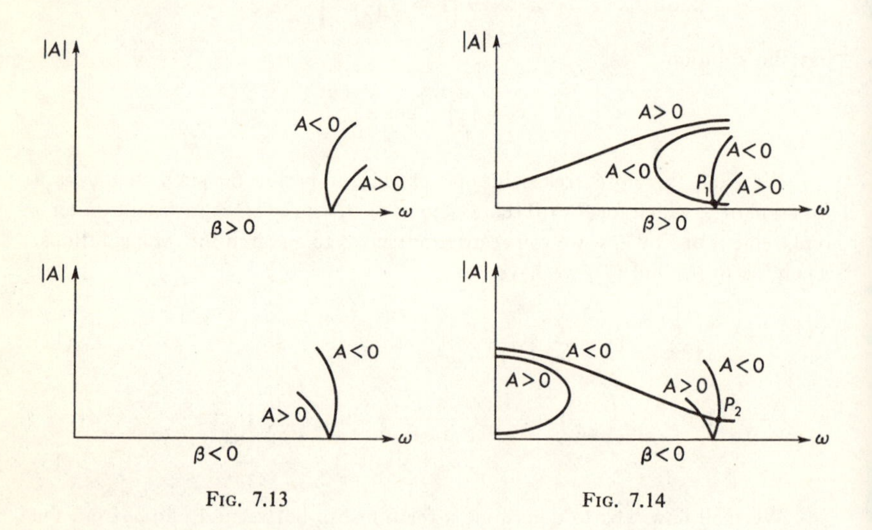

FIG. 7.13

FIG. 7.14

The points marked P_1 and P_2 represent so-called points of bifurcation. For the particular values of ω and A represented by these, we obtain a solution to our differential equation (11). If we now allow ω to vary, A may follow one or the other of the two possible response curves (10) or (15). In other words, a given solution may (upon a variation of the conditions of the problem) split, or bifurcate, into one of the two possible solutions.

As a last and final illustration of perturbation techniques for second-order systems, we turn to the autonomous equation:

$$x''+x=\beta(x'-\tfrac{1}{3}x'^3). \tag{16}$$

By differentiating (16) and letting $x' = y$, we obtain

$$y'' - \beta(1 - y^2)y' + y = 0.$$

It is clear from the Levinson-Smith theorem that this equation has a unique limit cycle. We shall, however, investigate (16) directly for small β via perturbation techniques. First we rewrite (16) as an integral equation:

$$x = A \cos z + B \sin z + \beta \int_0^z \sin(z - s)\,(x' - \tfrac{1}{3}x'^3)\,ds, \tag{17}$$

$$x' = -A \sin z + B \cos z + \beta \int_0^z \cos(z - s)\,(x' - \tfrac{1}{3}x'^3)\,ds. \tag{18}$$

We wish to determine suitable initial conditions that yield periodic solutions. For $\beta = 0$, we expect the period 2π, but as was seen in the preceding chapter, in the case of autonomous equations we must expect a variation in period also. We therefore require of (17) and (18) that

$$x(A, B, \beta, 2\pi + \tau) = x(A, B, \beta, 0) = A, \tag{19}$$

$$x'(A, B, \beta, 2\pi + \tau) = x'(A, B, \beta, 0) = B, \tag{20}$$

where the explicit dependence of x on A, B and β is brought out, and τ represents the change in period. We shall now expand the left sides of (19) and (20) for small values of β and τ. Then, from (17) and (18), we obtain

$$A + B\tau - \beta \int_0^{2\pi} \sin s(x' - \tfrac{1}{3}x'^3)\,ds + \cdots = A,$$

$$-A\tau + B + \beta \int_0^{2\pi} \cos s(x' - \tfrac{1}{3}x'^3)\,ds + \cdots = B.$$

For small β we can replace x' in the above integrals by

$$x' = -A \sin s + B \cos s.$$

After performing the required integrations, we have

$$\begin{aligned} \frac{\tau}{\pi\beta} B &= -A + \frac{1}{4} A(A^2 + B^2), \\ \frac{\tau}{\pi\beta} A &= B - \frac{1}{4} B(A^2 + B^2). \end{aligned} \tag{21}$$

Upon elimination of τ from these two equations, we are left with

$$A^2 + B^2 - \tfrac{1}{4}(A^2 + B^2)^2 = 0.$$

This shows that one of the following two conditions must hold:

$$A^2 + B^2 = 0,$$
$$A^2 + B^2 = 4.$$

For the first of these we obtain only a vanishing solution. The second, when inserted in (21), shows that

$$\tau = 0.$$

It follows that for a suitable angle θ, we can write

$$A = 2 \cos \theta,$$
$$B = 2 \sin \theta.$$

From (17) we now have

$$x = 2 \cos (z - \theta) + \cdots. \tag{22}$$

The dots represent small terms. The amplitude of the leading term is 2, which shows that we have a limit cycle. No other initial amplitude would lead to a non-vanishing periodic solution. θ is arbitrary, but as pointed out earlier, limit cycles of autonomous differential equations are unique only to within translations in time.

The methods employed in this section to establish the existence of periodic solutions for certain equations of type (1) for small β follow the general treatment of Chapter 6. But they are clearly very difficult to apply, and only first approximation terms for small β are obtained. It is not at all clear how such first approximations can be improved. The general case (1) can be discussed with relatively mild restrictions on the right side. But in many problems much more can be said. Often one encounters problems in which β enters in an analytic fashion and so do the other quantities. We shall now return to (5) and assume that

$$\omega = 1 + \beta\omega_1 + \beta^2\omega_2 + \cdots,$$
$$x = x_0 + \beta x_1 + \beta^2 x_2 + \cdots.$$

If these are inserted in (5), one finds

$$\begin{aligned}(1 + 2\beta\omega_1 + \beta^2(\omega_1^2 + 2\omega_2) + \cdots)(x_0'' + \beta x_1'' + \beta^2 x_2'' + \cdots)\\ + (x_0 + \beta x_1 + \beta^2 x_2 + \cdots) + \beta(x_0^3 + 3\beta^2 x_0^2 x_1 + \cdots) = \beta F \cos s.\end{aligned} \tag{23}$$

By equating coefficients of corresponding powers of β, one obtains the following recurrence differential equations:

$$\begin{aligned} x_0'' + x_0 &= 0,\\ x_1'' + x_1 &= F \cos s - 2\omega_1 x_0'' - x_0^3,\\ x_2'' + x_2 &= -2\omega_1 x_1'' - (\omega_1^2 + 2\omega_2)\, x_0'' - 3x_0^2 x_1,\\ &\vdots \end{aligned} \tag{24}$$

Since the initial conditions are to be independent of β, we now let

$$x_0(0) = A, \qquad x_0'(0) = 0,$$

$$x_n(0) = 0, \qquad x_n'(0) = 0, \qquad n \geqslant 1.$$

Then we find from the first of the equations (24) that

$$x_0 = A \cos s,$$

and using that in the second equation, we find

$$\begin{aligned} x_1'' + x_1 &= F \cos s + 2\omega_1 A \cos s - A^3 \cos^3 s \\ &= \left(F + 2\omega_1 A - \frac{3}{4} A^3\right) \cos s - \frac{1}{4} A^3 \cos 3s, \end{aligned}$$

$$x_1 = \frac{(F + 2\omega_1 A - \frac{3}{4} A^3)}{2} s \sin s + \frac{A^3}{32} (\cos 3s - \cos s). \tag{25}$$

So far, ω_1 is still unknown, but we are looking for periodic solutions. In general (25) cannot be expected to be periodic because it contains the term $s \sin s$. But we will now select ω_1 so that (25) is periodic; that is, the coefficient of $s \sin s$ must vanish. We thus find that

$$\omega_1 = \frac{3}{8} A^2 - \frac{F}{2A},$$

which was already obtained in (10). Since we know x_0, x_1, ω_1, we can now go on to the third equation in (24) and then solve for ω_2 by requiring that x_2 be periodic. These steps will not be carried out, but clearly the procedure can be used to obtain as many terms as are required.

The equation (16) can be approached in the same way, but it will be convenient in (16) to make a substitution for z. Let

$$z = \omega s,$$

so that (16) becomes

$$\omega^2 \frac{d^2x}{ds^2} + x = \beta \left(\omega \frac{dx}{ds} - \frac{1}{3} \omega^3 \left(\frac{dx}{ds}\right)^3\right).$$

The advantage of this substitution is that we now seek solution of period 2π in s and let ω depend on β. The period in z in (16) would not be fixed, but would depend on β. With this substitution we can operate with a period that is independent of β.

We again replace x and ω by power series in β, and as before, obtain recurrence differential equations analogous to (24). Then

$$\begin{aligned} x_0'' + x_0 &= 0, \\ x_1'' + x_1 &= -2\omega_1 x_0'' + x_0' - \frac{1}{3} x_0'^3. \\ &\vdots \end{aligned} \tag{26}$$

From the first of these we find that

$$x_0 = A \cos (s - \theta),$$

where both A and θ are still unknown. The second equation in (26) now becomes

$$\begin{aligned} x_1'' + x_1 &= + 2\omega_1 A \cos (s - \theta) - A \sin (s - \theta) + \frac{1}{3} A^3 \sin^3 (s - \theta) \\ &= 2\omega_1 A \cos (s - \theta) + \left(\frac{1}{4} A^3 - A\right) \sin (s - \theta) + \frac{1}{12} A^3 \sin 3(s - \theta). \end{aligned}$$

In order to ensure the periodicity of x_1, we require that the coefficients of both $\cos (s - \theta)$ and $\sin (s - \theta)$ vanish. Then

$$\begin{aligned} 2\omega_1 A &= 0, \\ \tfrac{1}{4} A^3 - A &= 0. \end{aligned}$$

Clearly, $A = 0$ is a solution, but leads only to the solution $x \equiv 0$. The other alternative is

$$\begin{aligned} \omega_1 &= 0, \\ A^2 &= 4, \end{aligned}$$

which agrees with the results obtained from (21), and we again obtain the solution

$$x_0 = 2 \cos (s - \theta),$$

showing that the amplitude cannot be arbitrarily prescribed. Here again we could obtain higher-order approximations if they were required.

Problems

1. In Section 7.1 it was shown that the linear equation

$$x'' + x = 0$$

can be solved in two steps. This involves solving the equation

$$\left(\frac{dx}{dz}\right)^2 + x^2 = c^2.$$

The former equation has unique solutions, whereas the latter does not. How do you reconcile these two statements? Does every solution of the latter satisfy the former?

2. In Section 7.1 the equations

$$X' = \begin{pmatrix} a & b \\ c & d \end{pmatrix} X$$

was discussed. To facilitate the work, A was diagonalized and the reduced equations were solved. Then certain curves were sketched. Show what these lóok like in terms of the original variables.

3. Show that

$$x'' + F(x') + x = 0,$$

where $F(y)$ is continuous, odd,

$$\begin{aligned} F(y) &< 0, \qquad \text{for } 0 < y < \bar{a}, \\ F(y) &> 0, \qquad \text{for } \bar{a} < y, \end{aligned}$$

and $F(y)$ grows monotonically to infinity for $y > \bar{a}$, has a unique, stable limit cycle.

4. Show that all solutions of

$$x'' + cx'^n + x = 0, \qquad c > 0,$$

are periodic for positive, even n, and sufficiently small initial values and that $x = 0$ is an asymptotically stable solution for positive, odd n.

5. Discuss the solutions of

$$\begin{aligned} x'' + f(x') + x &= 0, \\ f(x') &= -k, \qquad 0 < x' < v_1, \quad x' < -v_1, \\ &= k, \qquad v_1 < x', \qquad -v_1 < x' < 0. \end{aligned}$$

Do limit cycles exist?

6. Deduce the special case of (19) of section 7.2 for an equation of the type

$$x'' = f(x, x')$$

with periodic solutions.

Apply this result to the van der Pol equation (16) of Section 7.4.

7. Use the preceding exercise to estimate the rate at which a neighboring solution approaches a stable limit cycle.

8. Show that the index of a spiral is 1.

9. Property 3 of the Poincaré index was proved only for the case of two singularities. Extend the argument to the case of n singularities.

10. Study the solutions of

$$x'' = 4x - 5x^3 + x^5$$

via the methods of Section 7.3.

11. Apply the perturbation technique to (11)$'$ of Section 7.4 and show that the leading terms of the exact solution are obtained.

12. Study the solutions of the inhomogeneous van der Pol equations

$$x'' - \beta\left(x' - \frac{1}{3}x'^3\right) + X = \beta F \cos \omega z$$

via the perturbation technique.

13. Study the solutions of

$$x'' + x = \beta f(x, x')$$

by the perturbation technique for the following cases:

(a) $f(x, x') = f(x) - cx'$.

(b) $f(x, x') = f(x')$, where $f(-x') = f(x')$.

14. Study the solution of

$$x'' + x = \beta f(x, x' z)$$

by the perturbation technique for the following cases:

(a) $f(x, x', z) = f(x') + F \cos \omega z,\ x'f(x') > 0$ for $x' > 0$.

(b) $f(x, x', z) = f(x) - cx' + F \cos \omega\, z$.

15. Consider the Mathieu equation

$$x'' + [\lambda - \beta \cos 2z]x = 0.$$

In Chapter 5 it was shown that if λ is one of an infinite set of eigenvalues, the Mathieu equation will have solutions of period π or 2π. Find the first five eigenvalues, as functions of β, for small β by the perturbation technique. For $\beta = 0$, they are given by $\lambda_0 = 0$, $\lambda_1 = \lambda_2 = 4$ for solutions of period π and $\lambda_1' = \lambda_2' = 1$ for solutions of period 2π.

Bibliography

Bellman, R., *Stability Theory of Differential Equations*. New York: McGraw-Hill Book Company, Inc., 1953.

Bieberbach, L., *Theorie der Differentialgleichungen*. Berlin: Springer, 1930.

——. *Theorie der gewöhnlichen Differentialgleichungen auf Funktiontheoretischer Grundlage dargestellt*. Berlin: Springer, 1953.

Birkhoff, G., and Rota, G., *Ordinary Differential Equations*. Boston: Ginn & Company, 1962.

Bôcher, M., *Leçons sur les méthodes de Sturm*. Paris: Gauthier-Villars, 1917.

Cesari, L., *Asymptotic Behavior and Stability Problems in Ordinary Differential Equations*, 2d ed. New York: Academic Press, 1963.

Coddington, E. A., and Levinson, N., *Theory of Ordinary Differential Equations* New York: McGraw-Hill Book Company, Inc., 1955.

Davis, H. T., *Nonlinear Differential and Integral Equations*. New York: Dover, 1961.

Erdelyi, A. (Editor). *Higher Transandental Functions* (three volumes). New York: McGraw-Hill Book Company, 1955.

Hochstadt, H., *Special Functions of Mathematical Physics*. New York: Holt, Rinehart and Winston, Inc., 1961.

Hurewicz, W., *Lectures on Ordinary Differential Equations*. New York: John Wiley & Sons, Inc., 1958.

Ince, E. L., *Ordinary Differential Equations*. New York: Dover, 1956.

Kamke, E., *Differentialgleichungen: Lösungsmethoden und Lösungen*. Leipzig: Akademische Verlagsgesellschaft, 1943.

——. *Differentialgleichungen reeller Funktionen*. Leipzig, Akademische Verlagsgesellschaft, 1930.

Kaplan, W., *Ordinary Differential Equations*. Reading, Mass: Addison-Wesley Publishing Company, Inc., 1958.

Lanczos, C., *Linear Differential Operators*. Princeton, New Jersey: Van Nostrand, 1961.

Lefschetz, S., *Differential Equations, Geometric Theory*, 2d ed. New York: Interscience Publishers, Inc.

Liapounoff, A. M., *Problème Générale de la Stabilité du Mouvement*. Annals of Mathematics Studies, No. 17. Princeton, New Jersey: Princeton University Press, 1947.

McLachlan, N. W., *Ordinary Non-Linear Differential Equations in Engineering and Physical Sciences*. Oxford: Clarendon, Press, 1950.

——. *Theory and Application of Mathieu Functions*. Oxford, Clarendon Press, 1947.

Martin, W. T., and Reissner, E., *Elementary Differential Equations*, 2d. ed. Reading, Mass: Addison-Wesley Publishing Company, Inc., 1961.

Meixner, J., and Schäfke, F. W., *Mathieusche Funktionen und Sphäroidfunktionen*. Berlin: Springer, 1954.

Minorsky, N., *Introduction to Non-Linear Mechanics*. Ann Arbor, Michigan: Edwards Brothers, 1947.

Moulton, F. R., *Differential Equations*. New York: Dover, 1958.

Niemytskii, V., and Stepanov, V., *Qualitative Theory of Differential Equations*. Princeton, New Jersey: Princeton University Press, 1959.

Poincaré, H., *Les Méthodes Nouvelles de la Mécanique Céleste*, Volumes 1-3. New York: Dover Publications, Inc., 1957.

Pontryagin, L. S., *Ordinary Differential Equations*. Reading, Mass.: Addison-Wesley Publishers Company, Inc., 1962.

Poole, E. G. C., *Introduction to the Theory of Linear Differential Equations*. Oxford: Clarendon Press, 1936.

Sansone, G., *Equazioni Differenziali nel Campo Reale*, Volumes 1 and 2. Bologna, Italy: Zanichelli, 1948.

——., and Conti, R., *Equazioni Differenziali Non-Lineari*. Rome: Cremonese, 1956.

Stoker, J. J. *Nonlinear Vibrations in Mechanical and Electrical Systems*. New York: Interscience Publishers, Inc., 1950.

Struble, R. A., *Nonlinear Differential Equations*. New York: McGraw-Hill Book Company, 1962.

Titchmarsh, E. C., *Eigenfunction Expansions Associated with Second-Order Differential Equations*, Volume 1. Oxford: Clarendon Press, 1946: Volume 2, 1950.

Tricomi, F. G., *Differential Equations*. New York: Hafner Publishing Company, 1961.

Yosida, K., *Lectures on Differential and Integral Equations*. New York: Interscience Publishers, Inc., 1960.

Index

A CATALOGUE OF SELECTED DOVER BOOKS
IN ALL FIELDS OF INTEREST

A MAYA GRAMMAR, Alfred M. Tozzer. Practical, useful English-language grammar by the Harvard anthropologist who was one of the three greatest American scholars in the area of Maya culture. Phonetics, grammatical processes, syntax, more. 301pp. 5⅜ x 8½. 23465-7 Pa. $4.00

THE JOURNAL OF HENRY D. THOREAU, edited by Bradford Torrey, F. H. Allen. Complete reprinting of 14 volumes, 1837-61, over two million words; the sourcebooks for *Walden*, etc. Definitive. All original sketches, plus 75 photographs. Introduction by Walter Harding. Total of 1804pp. 8½ x 12¼. 20312-3, 20313-1 Clothbd., Two-vol. set $50.00

CLASSIC GHOST STORIES, Charles Dickens and others. 18 wonderful stories you've wanted to reread: "The Monkey's Paw," "The House and the Brain," "The Upper Berth," "The Signalman," "Dracula's Guest," "The Tapestried Chamber," etc. Dickens, Scott, Mary Shelley, Stoker, etc. 330pp. 5⅜ x 8½. 20735-8 Pa. $3.50

SEVEN SCIENCE FICTION NOVELS, H. G. Wells. Full novels. *First Men in the Moon, Island of Dr. Moreau, War of the Worlds, Food of the Gods, Invisible Man, Time Machine, In the Days of the Comet.* A basic science-fiction library. 1015pp. 5⅜ x 8½. (Available in U.S. only)
20264-X Clothbd. $8.95

ARMADALE, Wilkie Collins. Third great mystery novel by the author of *The Woman in White* and *The Moonstone.* Ingeniously plotted narrative shows an exceptional command of character, incident and mood. Original magazine version with 40 illustrations. 597pp. 5⅜ x 8½.
23429-0 Pa. $5.00

MASTERS OF MYSTERY, H. Douglas Thomson. The first book in English (1931) devoted to history and aesthetics of detective story. Poe, Doyle, LeFanu, Dickens, many others, up to 1930. New introduction and notes by E. F. Bleiler. 288pp. 5⅜ x 8½. (Available in U.S. only)
23606-4 Pa. $4.00

FLATLAND, E. A. Abbott. Science-fiction classic explores life of 2-D being in 3-D world. Read also as introduction to thought about hyperspace. Introduction by Banesh Hoffmann. 16 illustrations. 103pp. 5⅜ x 8½.
20001-9 Pa. $1.50

THREE SUPERNATURAL NOVELS OF THE VICTORIAN PERIOD, edited, with an introduction, by E. F. Bleiler. Reprinted complete and unabridged, three great classics of the supernatural: *The Haunted Hotel* by Wilkie Collins, *The Haunted House at Latchford* by Mrs. J. H. Riddell, and *The Lost Stradivarious* by J. Meade Falkner. 325pp. 5⅜ x 8½.
22571-2 Pa. $4.00

AYESHA: THE RETURN OF "SHE," H. Rider Haggard. Virtuoso sequel featuring the great mythic creation, Ayesha, in an adventure that is fully as good as the first book, *She*. Original magazine version, with 47 original illustrations by Maurice Greiffenhagen. 189pp. 6½ x 9¼.
23649-8 Pa. $3.00

AMERICAN BIRD ENGRAVINGS, Alexander Wilson et al. All 76 plates. from Wilson's *American Ornithology* (1808-14), most important ornithological work before Audubon, plus 27 plates from the supplement (1825-33) by Charles Bonaparte. Over 250 birds portrayed. 8 plates also reproduced in full color. 111pp. 9⅜ x 12½. 23195-X Pa. $6.00

CRUICKSHANK'S PHOTOGRAPHS OF BIRDS OF AMERICA, Allan D. Cruickshank. Great ornithologist, photographer presents 177 closeups, groupings, panoramas, flightings, etc., of about 150 different birds. Expanded *Wings in the Wilderness.* Introduction by Helen G. Cruickshank. 191pp. 8¼ x 11. 23497-5 Pa. $6.00

AMERICAN WILDLIFE AND PLANTS, A. C. Martin, et al. Describes food habits of more than 1000 species of mammals, birds, fish. Special treatment of important food plants. Over 300 illustrations. 500pp. 5⅜ x 8½. 20793-5 Pa. $4.95

THE PEOPLE CALLED SHAKERS, Edward D. Andrews. Lifetime of research, definitive study of Shakers: origins, beliefs, practices, dances, social organization, furniture and crafts, impact on 19th-century USA, present heritage. Indispensable to student of American history, collector. 33 illustrations. 351pp. 5⅜ x 8½. 21081-2 Pa. $4.00

OLD NEW YORK IN EARLY PHOTOGRAPHS, Mary Black. New York City as it was in 1853-1901, through 196 wonderful photographs from N.-Y. Historical Society. Great Blizzard, Lincoln's funeral procession, great buildings. 228pp. 9 x 12. 22907-6 Pa. $7.95

MR. LINCOLN'S CAMERA MAN: MATHEW BRADY, Roy Meredith. Over 300 Brady photos reproduced directly from original negatives, photos. Jackson, Webster, Grant, Lee, Carnegie, Barnum; Lincoln; Battle Smoke, Death of Rebel Sniper, Atlanta Just After Capture. Lively commentary. 368pp. 8⅜ x 11¼. 23021-X Pa. $6.95

TRAVELS OF WILLIAM BARTRAM, William Bartram. From 1773-8, Bartram explored Northern Florida, Georgia, Carolinas, and reported on wild life, plants, Indians, early settlers. Basic account for period, entertaining reading. Edited by Mark Van Doren. 13 illustrations. 141pp. 5⅜ x 8½. 20013-2 Pa. $4.50

THE GENTLEMAN AND CABINET MAKER'S DIRECTOR, Thomas Chippendale. Full reprint, 1762 style book, most influential of all time; chairs, tables, sofas, mirrors, cabinets, etc. 200 plates, plus 24 photographs of surviving pieces. 249pp. 9⅞ x 12¾. 21601-2 Pa. $6.50

AMERICAN CARRIAGES, SLEIGHS, SULKIES AND CARTS, edited by Don H. Berkebile. 168 Victorian illustrations from catalogues, trade journals, fully captioned. Useful for artists. Author is Assoc. Curator, Div. of Transportation of Smithsonian Institution. 168pp. 8½ x 9½. 23328-6 Pa. $5.00

HISTORY OF BACTERIOLOGY, William Bulloch. The only comprehensive history of bacteriology from the beginnings through the 19th century. Special emphasis is given to biography-Leeuwenhoek, etc. Brief accounts of 350 bacteriologists form a separate section. No clearer, fuller study, suitable to scientists and general readers, has yet been written. 52 illustrations. 448pp. 5⅝ x 8¼. 23761-3 Pa. $6.50

THE COMPLETE NONSENSE OF EDWARD LEAR, Edward Lear. All nonsense limericks, zany alphabets, Owl and Pussycat, songs, nonsense botany, etc., illustrated by Lear. Total of 321pp. 5⅜ x 8½. (Available in U.S. only) 20167-8 Pa. $3.00

INGENIOUS MATHEMATICAL PROBLEMS AND METHODS, Louis A. Graham. Sophisticated material from Graham *Dial,* applied and pure; stresses solution methods. Logic, number theory, networks, inversions, etc. 237pp. 5⅜ x 8½. 20545-2 Pa. $3.50

BEST MATHEMATICAL PUZZLES OF SAM LOYD, edited by Martin Gardner. Bizarre, original, whimsical puzzles by America's greatest puzzler. From fabulously rare *Cyclopedia,* including famous 14-15 puzzles, the Horse of a Different Color, 115 more. Elementary math. 150 illustrations. 167pp. 5⅜ x 8½. 20498-7 Pa. $2.50

THE BASIS OF COMBINATION IN CHESS, J. du Mont. Easy-to-follow, instructive book on elements of combination play, with chapters on each piece and every powerful combination team—two knights, bishop and knight, rook and bishop, etc. 250 diagrams. 218pp. 5⅜ x 8½. (Available in U.S. only) 23644-7 Pa. $3.50

MODERN CHESS STRATEGY, Ludek Pachman. The use of the queen, the active king, exchanges, pawn play, the center, weak squares, etc. Section on rook alone worth price of the book. Stress on the moderns. Often considered the most important book on strategy. 314pp. 5⅜ x 8½. 20290-9 Pa. $3.50

LASKER'S MANUAL OF CHESS, Dr. Emanuel Lasker. Great world champion offers very thorough coverage of all aspects of chess. Combinations, position play, openings, end game, aesthetics of chess, philosophy of struggle, much more. Filled with analyzed games. 390pp. 5⅜ x 8½. 20640-8 Pa. $4.00

500 MASTER GAMES OF CHESS, S. Tartakower, J. du Mont. Vast collection of great chess games from 1798-1938, with much material nowhere else readily available. Fully annoted, arranged by opening for easier study. 664pp. 5⅜ x 8½. 23208-5 Pa. $6.00

A GUIDE TO CHESS ENDINGS, Dr. Max Euwe, David Hooper. One of the finest modern works on chess endings. Thorough analysis of the most frequently encountered endings by former world champion. 331 examples, each with diagram. 248pp. 5⅜ x 8½. 23332-4 Pa. $3.50

SECOND PIATIGORSKY CUP, edited by Isaac Kashdan. One of the greatest tournament books ever produced in the English language. All 90 games of the 1966 tournament, annotated by players, most annotated by both players. Features Petrosian, Spassky, Fischer, Larsen, six others. 228pp. 5⅜ x 8½. 23572-6 Pa. $3.50

ENCYCLOPEDIA OF CARD TRICKS, revised and edited by Jean Hugard. How to perform over 600 card tricks, devised by the world's greatest magicians: impromptus, spelling tricks, key cards, using special packs, much, much more. Additional chapter on card technique. 66 illustrations. 402pp. 5⅜ x 8½. (Available in U.S. only) 21252-1 Pa. $3.95

MAGIC: STAGE ILLUSIONS, SPECIAL EFFECTS AND TRICK PHOTOGRAPHY, Albert A. Hopkins, Henry R. Evans. One of the great classics; fullest, most authorative explanation of vanishing lady, levitations, scores of other great stage effects. Also small magic, automata, stunts. 446 illustrations. 556pp. 5⅜ x 8½. 23344-8 Pa. $5.00

THE SECRETS OF HOUDINI, J. C. Cannell. Classic study of Houdini's incredible magic, exposing closely-kept professional secrets and revealing, in general terms, the whole art of stage magic. 67 illustrations. 279pp. 5⅜ x 8½. 22913-0 Pa. $3.00

HOFFMANN'S MODERN MAGIC, Professor Hoffmann. One of the best, and best-known, magicians' manuals of the past century. Hundreds of tricks from card tricks and simple sleight of hand to elaborate illusions involving construction of complicated machinery. 332 illustrations. 563pp. 5⅜ x 8½. 23623-4 Pa. $6.00

MADAME PRUNIER'S FISH COOKERY BOOK, Mme. S. B. Prunier. More than 1000 recipes from world famous Prunier's of Paris and London, specially adapted here for American kitchen. Grilled tournedos with anchovy butter, Lobster a la Bordelaise, Prunier's prized desserts, more. Glossary. 340pp. 5⅜ x 8½. (Available in U.S. only) 22679-4 Pa. $3.00

FRENCH COUNTRY COOKING FOR AMERICANS, Louis Diat. 500 easy-to-make, authentic provincial recipes compiled by former head chef at New York's Fitz-Carlton Hotel: onion soup, lamb stew, potato pie, more. 309pp. 5⅜ x 8½. 23665-X Pa. $3.95

SAUCES, FRENCH AND FAMOUS, Louis Diat. Complete book gives over 200 specific recipes: bechamel, Bordelaise, hollandaise, Cumberland, apricot, etc. Author was one of this century's finest chefs, originator of vichyssoise and many other dishes. Index. 156pp. 5⅜ x 8. 23663-3 Pa. $2.50

TOLL HOUSE TRIED AND TRUE RECIPES, Ruth Graves Wakefield. Authentic recipes from the famous Mass. restaurant: popovers, veal and ham loaf, Toll House baked beans, chocolate cake crumb pudding, much more. Many helpful hints. Nearly 700 recipes. Index. 376pp. 5⅜ x 8½. 23560-2 Pa. $4.00

THE AMERICAN SENATOR, Anthony Trollope. Little known, long unavailable Trollope novel on a grand scale. Here are humorous comment on American vs. English culture, and stunning portrayal of a heroine/villainess. Superb evocation of Victorian village life. 561pp. 5⅜ x 8½.
23801-6 Pa. $6.00

WAS IT MURDER? James Hilton. The author of *Lost Horizon* and *Goodbye, Mr. Chips* wrote one detective novel (under a pen-name) which was quickly forgotten and virtually lost, even at the height of Hilton's fame. This edition brings it back—a finely crafted public school puzzle resplendent with Hilton's stylish atmosphere. A thoroughly English thriller by the creator of Shangri-la. 252pp. 5⅜ x 8. (Available in U.S. only)
23774-5 Pa. $3.00

CENTRAL PARK: A PHOTOGRAPHIC GUIDE, Victor Laredo and Henry Hope Reed. 121 superb photographs show dramatic views of Central Park: Bethesda Fountain, Cleopatra's Needle, Sheep Meadow, the Blockhouse, plus people engaged in many park activities: ice skating, bike riding, etc. Captions by former Curator of Central Park, Henry Hope Reed, provide historical view, changes, etc. Also photos of N.Y. landmarks on park's periphery. 96pp. 8½ x 11. 23750-8 Pa. $4.50

NANTUCKET IN THE NINETEENTH CENTURY, Clay Lancaster. 180 rare photographs, stereographs, maps, drawings and floor plans recreate unique American island society. Authentic scenes of shipwreck, lighthouses, streets, homes are arranged in geographic sequence to provide walking-tour guide to old Nantucket existing today. Introduction, captions. 160pp. 8⅞ x 11¾. 23747-8 Pa. $6.95

STONE AND MAN: A PHOTOGRAPHIC EXPLORATION, Andreas Feininger. 106 photographs by *Life* photographer Feininger portray man's deep passion for stone through the ages. Stonehenge-like megaliths, fortified towns, sculpted marble and crumbling tenements show textures, beauties, fascination. 128pp. 9¼ x 10¾. 23756-7 Pa. $5.95

CIRCLES, A MATHEMATICAL VIEW, D. Pedoe. Fundamental aspects of college geometry, non-Euclidean geometry, and other branches of mathematics: representing circle by point. Poincare model, isoperimetric property, etc. Stimulating recreational reading. 66 figures. 96pp. 5⅝ x 8¼.
63698-4 Pa. $2.75

THE DISCOVERY OF NEPTUNE, Morton Grosser. Dramatic scientific history of the investigations leading up to the actual discovery of the eighth planet of our solar system. Lucid, well-researched book by well-known historian of science. 172pp. 5⅜ x 8½. 23726-5 Pa. $3.00

THE DEVIL'S DICTIONARY. Ambrose Bierce. Barbed, bitter, brilliant witticisms in the form of a dictionary. Best, most ferocious satire America has produced. 145pp. 5⅜ x 8½. 20487-1 Pa. $1.75